Creating Scientific Controversies

For decades, cigarette companies helped to promote the impression that there was no scientific consensus concerning the safety of their product. The appearance of controversy, however, was misleading, designed to confuse the public and to protect industry interests. Created scientific controversies emerge when expert communities are in broad agreement but the public perception is one of profound scientific uncertainty and doubt. In the first book-length analysis of the concept of a created scientific controversy, David Harker explores issues including climate change, Creation science, the anti-vaccine movement and genetically modified crops. Drawing on work in cognitive psychology, social epistemology, critical thinking and philosophy of science, he shows readers how to better understand, evaluate and respond to the appearance of scientific controversy. His book will be a valuable resource for students of philosophy of science, environmental and health sciences, and social and natural sciences.

David Harker is Associate Professor of Philosophy at East Tennessee State University. He has published articles in journals including *British Journal for the Philosophy of Science, Philosophical Studies* and *Studies in History and Philosophy of Science*.

Creating Scientific Controversies

Uncertainty and Bias in Science and Society

David Harker

Cambridge University Press

CAMBRIDGE
UNIVERSITY PRESS

University Printing House, Cambridge CB2 8BS, United Kingdom

Cambridge University Press is part of the University of Cambridge.

It furthers the University's mission by disseminating knowledge in the pursuit of education, learning and research at the highest international levels of excellence.

www.cambridge.org
Information on this title: www.cambridge.org/9781107692367

First published 2015

Printed in the United Kingdom by TJ International Ltd. Padstow Cornwall

A catalogue record for this publication is available from the British Library

Library of Congress Cataloguing in Publication data
Harker, David (David W.)
Creating scientific controversies : uncertainty and bias in science and society / David Harker.
pages cm
Includes bibliographical references and index.
ISBN 978-1-107-69236-7
1. Science – Philosophy. 2. Skepticism. I. Title.
Q175.H3245 2015
501–dc23
2015011610

ISBN 978-1-107-06961-9 Hardback
ISBN 978-1-107-69236-7 Paperback

For Harry and Isla, in the hopes that someday they might choose to read it, and then review it favourably

Contents

Preface

For several years I have enjoyed teaching an introductory college course that explored many of the issues discussed in this book. My thanks go out to all those students who have helped me think through these issues, who have challenged me to find better ways of explaining the material, and who have helped me see which aspects were of greater or lesser relevance. Several friends and colleagues were extremely generous with their time and talents, reading through large sections of the book, and providing wonderful feedback that greatly improved the book. Thanks in particular to Bob Schroer, Justin Sytsma, Matt Lund, Dave Hilbert, Nick Huggett and Bob Fischer. Parts of the book were presented to audiences in Durham, Leeds and Bristol. I am very grateful to those who made these events possible, and to those who attended and offered helpful questions and discussion. An anonymous referee from Cambridge University Press made several excellent suggestions that I'm sure have made the book better. My editors, Hilary Gaskin and Rosemary Crawley, were incredibly helpful with the book's preparation, for which I am very grateful. My thanks are also owed to the College of Arts and Science at East Tennessee State University, for awarding me a Summer Research Fellowship in 2014, which helped in the final stages of writing. My sister and parents have always provided enormous encouragement and support. There are many reasons why this book wouldn't have been written if it wasn't for them. Finally, my wife is a perennial source of inspiration and optimism. With respect to this project she was always willing to offer support, advice and reassurance. For the many ways she enriches my life I am indebted.

Introduction: scientific authority and the created controversy

The word *science* can conjure up for us a variety of ideas and images. It can whisk us back to cluttered classrooms, furnished with tall stools and long benches, Bunsen burners, and glass-doored cupboards stocked with assorted paraphernalia. The word might bring to mind the names and faces of famous scientists: Newton, Faraday, Hawking, Curie, Galileo and Mendel all jostle for attention, but ultimately are crowded out by a mental image of Freud and his cigar, Einstein and his untamed hair, or Darwin and an almost equally untamed beard. Maybe it rouses important scientific concepts, activities or instruments, the atom, star gazing, the test tube or the microscope. Perhaps we imagine a pristine laboratory, a young technician dressed in an immaculate white coat, scrutinizing a vial of blue translucent fluid for reasons unknown.

Science is ubiquitous. Its boundaries are fuzzy, its range bewildering. Distinctions have been drawn between different *types* of science, natural versus social, hard versus soft, historical versus experimental, and so on. Disagreement reigns over whether economics is science, whether anthropology is science, whether history is science. Creation science calls itself science, but many call foul. Politicians have suggested – what sounds thoroughly reasonable – that policy should utilize *sound science* and eschew *junk science*. Scientific discoveries are reported in the media; scientific concepts are utilized in novels, film and television. Science is popularized and demonized. It offers explanations of our most commonplace observations, but in terms that are peculiar and hard to comprehend. Scientific developments are integral to some of society's most remarkable achievements, but also some of our most horrifying tragedies. Science is both utterly familiar and an immediate source of controversy and debate.

The fact that science occupies an extraordinarily important place within modern society makes it important to think more carefully about what science is. A great deal of scientific research is conducted in service to issues of public safety and perceived public need, is funded by taxpayers, and is overseen at least to some extent by political systems. It is sensible to consider whether the research being pursued is of genuine value, of greater value than research that doesn't get funded, and whether the degree of political oversight is appropriate. The technological fruits of scientific labours often give rise to hard questions about the ethics of warfare, human reproduction, food production, energy development and more. To some extent at least science interacts with every other aspect of society, and at each point of interaction there is room to evaluate whether sensible goals have been identified, and whether sensible methods are being deployed with respect to those goals.

The *applications* of scientific research generate critical questions, but even if we restrict our attention to science as it seems principally concerned with generating facts, or knowledge, or information, our preconceptions deserve closer scrutiny. Insofar as the laboratory white coat seems emblematic of science, do we forget those scientists who work predominantly in the field, or at a computer, or with human subjects? Do we regard some of these activities as *less scientific* and, if so, on what grounds? Newton and Galileo are famous figures from the history of science, but how sure are we that today's scientists would agree with them about what counts as science, and how science should be conducted? Distinguishing scientific from non-scientific disciplines sounds like a sensible and worthwhile endeavour, but significant efforts have accomplished little consensus among philosophers of science; neither the struggles, with what's known as the problem of demarcation, nor their implications, are widely appreciated. Thus, even ignoring the broader roles of science in the modern world, attempts to understand the nature of science – its methods, assumptions, limitations and achievements – prompt difficult questions and sensible concerns. Making sense of even this much is important, as we'll see, but not straightforward (as we'll also see).

Regardless of where precisely we define or find its edges, however, and what exactly we imagine happens between them, most of us appear, at least most of the time, willing to admit science as our most reliable means of acquiring knowledge of both ourselves and the world we inhabit. Scientists are our most respected authorities on an incredibly wide variety of issues. We look to them for answers concerning the deep past and the near future, the living and the inanimate, the farthest reaches of the universe and the innermost secrets of the human mind. An enormously diverse range of subjects are lumped together under the banner of science, and one of the few attributes these subjects share in common is that *because they're science* we attach special significance to their conclusions. We assume science is more *objective* than alternative ways of investigating the world, that it is guided by *facts* and thus less influenced by fads and impulse, its results unvarnished by personal agenda, prejudice and bias. By overcoming biases science is considered more *rational* than alternative ways of investigating, and thereby achieves something that is absent from many other human activities: the histories of literature, art, and music contain ample evidence of change, but science seems to do more – it makes *progress*. Our confidence in its methods means we trust science to tell us how things are, at least most of the time. The reason it is sometimes thought important whether psychology, for example, can properly be called a *science* is that the term has significant rhetorical force; merely labelling a discipline *scientific* can improve its profile and credentials, hence the occasionally bitter disputes over whether a particular discipline deserves the honorific.

The confidence we place in science deserves close inspection. Can we justify our preference for distinctively scientific conclusions? Can we identify those features that make science more rational and more objective? As we'll see, that science is purportedly more objective, more rational, based on facts, and so on are ideas that have all been challenged. In the early 1960s a fresh approach to questions about the nature of science revolutionized philosophy of science and neighbouring disciplines, and influenced many more distantly related fields. Some who were stirred by the new methodology advanced yet more radical challenges during the 1970s and 1980s, which ultimately led to what the media dramatically dubbed *The Science Wars*. What appeared to be at stake was nothing less than scientific authority itself – that widespread assumption which admits science as our most reliable means of generating knowledge. Many would defend a triumphal, laudatory attitude towards science, but others were far less sanguine. In subsequent chapters, we'll discuss, in broad outline, both the evolution of these ideas and their foundations.

Despite the influence of these sceptical attitudes towards science, in many circles, and in many contexts, scientific authority remains unsullied. Even putting radically sceptical attitudes towards science aside, however, the authority of science can still seem somewhat surprising if we pause to reflect on some modern scientific conclusions. Physics and chemistry trade in objects that seem almost impossibly small. Consider that there are more atoms in a grain of sand than there are grains of sand on almost any given beach. (Read that sentence again if at first glance it seemed just too incredible.) Cosmology and geology deal with events on enormous scales, both in space and time, and processes that take so long to unfold that the entire history of mankind is negligible by comparison. Compress the history of Earth into one calendar year and the birth of Jesus Christ occurs with less than fifteen seconds of the year remaining. (You might try pondering on that when you're about to begin the countdown next New Year's Eve.) Molecular biologists tell us that the information needed to create something as complicated as a human being is contained within twenty-three pairs of chromosomes, that two copies of each chromosome can be found in almost every cell in your body, but most of these cells are so small that they're invisible to the naked eye. Einstein's theories of special and general relativity, quantum physics, neuroscience, biochemistry, evolutionary biology, and a great deal more besides, can simultaneously stupefy and delight. Science presents us with a fascinating description of the world, often highly unexpected, and sometimes so mind bogglingly bizarre that it stretches credulity.

Yet in spite of its astonishing conclusions most of us appear willing to accept most scientific claims without fuss, perhaps through familiarity, or a tendency to massage the claims into a more digestible form. The truly remarkable nature of some of science's most important discoveries can be easily overlooked, but should serve to illustrate that our personal intuitions, speculations and extrapolations from familiar experiences, our common-sense judgements, and even just our very best efforts to think really hard, rarely produce results that even remotely approximate the scientific image. Conceding the authority of science requires that we relinquish the right to reject scientific conclusions just because they don't sit right with us. If our gut reactions were a reliable guide to the plausibility of scientific conclusions, then scientific successes wouldn't keep building on ideas that at first sound preposterous. That our planet spins on its axis at hundreds of miles an hour, and completes an annual orbit of the Sun, were deeply disconcerting ideas to Galileo's contemporaries in the early seventeenth century. The size of the universe that was needed by the heliocentric model, to account for the absence of what's known as stellar parallax, was nothing short of astounding.[1] The discomfort felt by those who were challenged by Galileo was in itself an unreliable reason for rejecting heliocentricism. Mere discomfort isn't a better reason for rejecting a scientific conclusion today.

Nevertheless, on a wide range of issues many people are accused of dismissing well-established scientific claims, despite the absence of any good reason for doing so. Advocates for certain scientific ideas insist that the science is now beyond dispute, and that its main conclusions can't be ignored. Those who continue to resist these conclusions are accused of relying on superstition, or wishful thinking, or simply of being ignorant and confused. Science proponents insist that too many people, without relevant qualifications or understanding, trust their own ability to evaluate the state of the science above the combined acumen of scores of experts. And the result is more serious than simply a misinformed public. By ignoring our scientists, it is suggested, we are risking our health and the health of our children, flirting with unprecedented environmental disaster, and thereby incalculable human cost, embracing naïve attitudes about the world around us, denying ourselves technological advances that hold the potential for momentous improvements to our quality of life and unnecessarily endangering members of society that deserve far greater protection. Throughout this book I'll assume that such consequences are bad and that we should be motivated to

[1] It's familiar from everyday experience that objects change their apparent position, relative to more distant objects, as we change our perspective. For example, with just one eye open, hold your thumb out at arm's length so that it fully obstructs your view of some object. Now switch eyes and the object becomes visible, with your thumb located some distance to the left or right of that same object within your visual field. The change in perspective affects the apparent position of your thumb, relative to the more distant object. If the Earth orbits the Sun, then our perspective on any given star changes. Stars that are closer should, therefore, appear to change their positions, relative to more distant stars. These effects are now discernible, but the technologies available to Galileo were insufficient to reveal this effect. Heliocentricism was reconciled with this failure to observe stellar parallax by supposing that even the closest stars were much farther from Earth than anyone had previously believed. The change in perspective, as the Earth orbits the Sun, thus becomes negligible relative to the distance to the stars, and hence the absence of stellar parallax is explained.

avoid them. Since I imagine that most readers would agree on both scores, the more important questions become: Are we really ignoring well-established scientific opinion without good reason, and, if so, how is this happening? How sure are we that science has got it right? Science has been wrong before, so why should we trust what today's scientists say? Furthermore, with respect to many scientifically informed, hot-button issues, like climate change, homeopathy, stem cell research, evolutionary biology, the safety of vaccinations, and so on, we are often aware of objections to the science, and to the presence of dissenting voices seemingly from within the relevant scientific community. In some instances, we might suspect that certain scientists have an agenda to promote their ideas regardless of the quantity and quality of the evidence that's available to them. Although some insist the science is unassailable, isn't it reasonable to still have doubts and concerns about even the most basic conclusions?

The stakes are high. Needlessly risking our health, our lives, and our planet, is something we all wish to avoid, but if the science is wrong, then endorsing its recommendations will also have avoidable and perhaps devastating consequences. We must strive to make decisions, whether on a personal, regional or global level, that are based on the best information we have, and hence we can afford to ignore prevailing scientific conclusions only if we have good reasons for doing so. In this book, we'll be concerned to learn how we can better evaluate the state of debates that are advertised as scientifically controversial, and to see what kinds of objections to scientific controversy are worth worrying about. In part this is a book about how we have been led astray on certain issues, and what we can do to avoid being misled again. We'll seek to improve our critical thinking skills and better appreciate some of our shortcomings as rational agents. We'll look to philosophy of science for help in developing more sensible attitudes towards the nature of scientific inquiry and, in the final chapters of the book, towards particular issues surrounding matters of public health, as well as climate change and Creation science. Reviewing an example now, however, will help better illustrate some of the book's principal themes; a particularly instructive example concerns the conduct of the tobacco companies during the middle of the twentieth century.[2]

The cigarette deception

The first machine-rolled, modern cigarettes appeared in the 1880s. Almost from their inception there were concerns about possible health risks. The effects of smoking on circulation, physical and mental development, as well as fertility and lactation in women, were all explored during the first few decades of the twentieth century, but

[2] The actions of the cigarette industry during the second half of the twentieth century are very well documented. Brandt (2009) and Proctor (2012) are excellent resources.

studies produced only ambiguous results. Excessive smoking was generally judged inadvisable, although it was unclear exactly what qualified as excessive. Smoking was also widely regarded as harmful to children, though this was seemingly based on little more than suspicion. Largely, however, medical opinion settled on moderate smoking as a safe practice for most adults. The fact that cigarettes increase the risks of developing lung cancer owed its discovery to several factors. Allan Brandt, a historian of medicine, argues that one of the most important factors was the introduction of new statistical methods into medical research during the late 1940s, methods that were introduced by the very researchers who were curious about the cigarette-lung cancer connection.

During the early decades of the twentieth century, the study of disease was dominated by laboratory-based techniques. The German physician, Robert Koch, had advanced four principles for identifying the causes of infectious disease. If a certain microorganism is found predominantly in organisms diagnosed with a given disease, and can then be isolated from the diseased organism and grown in culture, utilized to induce disease in an otherwise healthy organism, and, finally, isolated from the inoculated organism and shown to be identical with the original microorganism, then we can conclude that this particular type of microorganism causes the particular disease. Koch's postulates were hugely influential, although recognized even by Koch as having significant limitations; they also played an important role in the more general, dominant penchant for studying disease at the cellular level. An important implication of this preference for laboratory methods was the marginalization of statistics within medical research.

In 1950 two important papers were published, one by American researchers, Ernst Wynder and Evarts Graham, and a second by British scientists, Bradford Hill and Richard Doll.[3] These presented impressive statistical evidence relating cigarette smoking to lung cancer. Within their sample populations, lung cancer was extremely rare among non-smokers, and cigarette use was high among those diagnosed with lung cancer. Among lung cancer patients, the high ratio of males to females was attributed to the fact that smoking had been a predominantly male activity. The evidence was, however, as the authors were only too aware, *merely* statistical. The use of statistics is by now so common that it's hard to appreciate how things could ever have been otherwise. Nevertheless, within the medical profession, overcoming scepticism and resistance towards the connection between cigarettes and lung cancer involved, in large part, convincing the medical community that statistics could provide a legitimate form of scientific evidence.

What's easy to overlook, in our efforts to better understand the nature of scientific inquiry, is that some scientific debates concern the propriety of particular *methods*.

[3] In the 1930s German and Argentine scientists had gathered evidence that cigarettes cause lung cancer, both through animal experimentation and statistical studies, but these were largely ignored. See Proctor (2012) for more details.

This is a failing that has important implications for our understanding of scientific controversies. For example, if scientific communities need persuading of the cogency of new approaches to traditional questions, then it is incumbent on us to learn more about those methods before we can justifiably dismiss the resultant conclusions. Second, cases like the cigarette controversy speak against the idea that science employs an unchanging set of methods: the ways in which science generates knowledge are themselves prone to revision. If scientific methods change, then what constitutes *good science* must also change. Establishing the connection between cigarettes and lung cancer involved an expansion of what qualified as good medical research. If the standards by which we evaluate science are susceptible to change, however, then an important incentive for distinguishing scientific from non-scientific research dissipates. In Chapter 1, we'll return to this idea when we consider the infamous problem of demarcation, and ask whether attempts to resolve the problem are a worthwhile exercise.

Returning to our historical narrative, in 1950 a number of prominent, qualified researchers were unconvinced by the initial statistical studies, but more studies followed and the evidence mounted. Lung cancer rates didn't vary between rural and urban areas, undermining the idea that air pollution was responsible for the rise in reported cases of lung cancer. Laboratory studies on mice corroborated the idea that something in cigarette smoke was carcinogenic. By the mid-1950s there were very few qualified experts denying that there was an important causal connection between cigarette smoking and lung cancer. The U.S. surgeon general was convinced. Leading medical associations were convinced. The response of the tobacco industry was swift, shrewd and hugely formative when it came to shaping *public* perceptions.

Cigarette companies were already tremendously adept at influencing public attitudes towards their product, having developed a variety of innovative, and highly effective, marketing strategies and advertising campaigns. (Cigarette companies are often credited with having essentially launched the marketing and advertising industries.) Such skills were now utilized and refined in response to the new science-based threat. Early advertising campaigns encouraged smokers to conduct their own personal tests and to see for themselves that a certain brand didn't cause throat irritation. Individuals were thereby urged to trust their own judgements over the science. With hindsight, we recognize the folly of those who thought that their personal experience with cigarettes could provide more reliable evidence than large population-based studies, but we will see later that the habit of dismissing science in favour of personal experience persists.

Fairly quickly, however, the strategies of the tobacco industry shifted. Rather than seek to undermine *science*, and promote confidence in personal experience, the cigarette companies sought to undermine the idea that there existed a scientific *consensus*. Rather than distance itself from science, and criticize scientific research from afar, the industry sought ways to achieve sufficient scientific credibility that they could corrupt from within. The industry gambled on the idea that creating the appearance of uncertainty within the scientific community would provide smokers with sufficient reason to continue smoking and many non-smokers insufficient reason not to start. If tobacco could successfully infiltrate the scientific process, then people could no longer look to science for answers; tobacco would create the impression that science had no answers, but offered only confusion and uncertainty. As one tobacco industry CEO would famously remark, 'Doubt is our product.' Uncertainty and doubt play a significant role within the kinds of debates we'll discuss later in the book. The absence of certainty surrounding a given conclusion is quite consistent with having very compelling evidence for that same conclusion. In certain circumstances, however, we can all display a strange tendency towards supposing that if no-one knows for *certain*, then no-one knows at all, and hence that any opinion on a given issue is as sensible as any other. Clearly we must learn to respond more responsibly to the appearance of doubt.

A 1953 statement, released on behalf of the major cigarette manufacturers, described their belief that cigarettes were not harmful, and their commitment to continue ensuring rigorous safety standards. Funds were promised by the industry, and were forthcoming, for additional research. Several decades on and it's clear that the projects which received industry funding were far more concerned with genetic and environmental explanations for lung cancer, and thus possessed greater potential to reprieve cigarettes than to generate additional evidence of blame. More important, by funding scientific research, even research over which they had a large degree of control, the tobacco industry bought itself a measure of scientific legitimacy, which could then be leveraged to create the appearance of a genuine scientific debate. Keeping the debate alive kept uncertainty alive. Keeping uncertainty alive helped sustain the tobacco companies' agenda.

The industry's scepticism was dressed up to fit familiar scientific virtues: according to the cigarette companies it was too early to offer definite conclusions about the connection; more studies were needed, as well as calm, careful analyses of available evidence. Such cautionary attitudes and requests for more research can sound eminently reasonable, by science's own standards. Those who were insisting that cigarettes were responsible for the increase in reported cases of lung cancer were portrayed as alarmist and irrational. Those scientists who disputed the majority view of health professionals were few in number, but the industry worked hard to publicize their opinions and puff their reputations. In general, as long as some scientists were willing to repeatedly dispute the connection between cigarettes and lung cancer – no matter who paid their salary – the public would hear that some scientists were convinced by the evidence and some were not.

The story of tobacco companies' deception is a familiar one, but there are morals we should be mindful of. A rallying cry of industry scientists was that there was *no proof* that cigarettes caused lung cancer. As we'll see in Chapter 2, however, we must be careful with our use of the word *proof* when discussing scientific claims; observing that a given conclusion has not been proven appears noteworthy only because of a subtle ambiguity in language, which reflecting on the nature of science can quickly dispel. A second conspicuous strategy of industry scientists drew attention to issues that weren't understood, with the hope of promoting uncertainty more broadly. What prompts people to start smoking? Why do some smoke more heavily than others? Why do many smokers suffer no ill health? Until we could remove such uncertainties, it was argued, it is imprudent to draw any conclusions about the state of the debate. But the argument is unpersuasive: we can achieve very good evidence that smoking causes lung cancer without having a good explanation for why people become smokers in the first place. Emphasizing what's not well understood by science is an effective means of *magnifying uncertainty*. By magnifying uncertainty groups can undermine public confidence in a much broader set of ideas. We will inspect the strategy more closely in Chapter 7.

Industry spokespeople employed double standards, according to whether studies spoke for or against the risks of smoking: evidence for the connection between smoking and lung cancer were held to impossibly high standards, but the idea that smoking is therapeutic was promoted without any serious evidence whatsoever; animal studies that suggested cigarettes were harmful were dismissed as uninformative on the grounds that what's true for mice is an unreliable guide to what's true for humans, but when animal studies purportedly provided favourable evidence for the industry, these were touted as important and telling. On behalf of cigarette companies, it was often noted that non-smokers did sometimes develop lung cancer and that many individuals smoked heavily throughout their lives seemingly without consequence. How, it was argued, could lung cancer thus be attributed to smoking? The argument commits a classic informal fallacy, known as the straw man fallacy. No-one claimed that all and only smokers would develop lung cancer, only that smoking increased the risks. We'll look at a variety of such fallacies in Chapter 6.

The industry promoted its own scientists, ideas and arguments, aggressively and persistently. For every article, report and editorial sympathetic to the mainstream wisdom that smoking caused lung cancer, the industry ensured that there appeared a rebuttal. The state of the science was often misrepresented: that cigarettes cause cancer was criticized for relying on merely statistical data, even though notable evidence from pathological and experimental research had been provided. Many of these tactics exploit important mechanisms by which we all evaluate claims and conclusions. The wishful thinking fallacy provides a plausible explanation for differing attitudes towards the risks of smoking, as held by smokers versus non-smokers. A 1991 Gallup poll reported that 91 per cent of non-smokers accepted smoking as a cause of lung cancer, compared with just 69 per cent of smokers.[4] Understanding more

[4] If this is an example of wishful thinking, then the diverging responses arise because people's desires and hopes are influencing what they believe. An alternative explanation for the polling results suggests instead that people who are

about our cognitive tendencies, and how they can sometimes lead us astray, can facilitate better evaluations of available evidences. We will briefly survey some recent work in cognitive psychology in Chapter 5.

By the mid-1950s, the medical community was in little doubt about the risks of smoking cigarettes, but cigarette companies continued to magnify uncertainty and doubt. According to a 1954 Gallup poll, 90 per cent of those surveyed were aware of reports that cigarettes may be a leading cause of lung cancer but, even by 1960, only 50 per cent accepted that cigarettes do cause cancer. It would be easy to feel smug when we reflect on all those who failed to recognize what by now seems so obvious, to shake our heads in disbelief that so many were so atrociously duped. However, we must be careful not to underestimate the enormous difficulties that were overcome by those who rendered this fact scientifically credible. We should be just as careful not to underestimate the obstacles overcome by those who convinced a sceptical public, obstacles that were largely erected and maintained by the remarkably well-funded, savvy and determined tobacco industry.

There are unquestionably a variety of factors – psychological, social and political – responsible for the success of the cigarette in the second half of the twentieth century, in spite of the incontrovertible evidence that described its grave health risks. Part of the explanation, however, is the public perception that the science was unsettled. There are many who would argue that the public is being similarly fooled, on a variety of issues, by the same tricks that tobacco did so much to develop. Consequently, there are important lessons to learn from the cigarette deception, lessons that can help us distinguish genuine scientific controversies from the mere appearance of controversy, lessons in critical thinking and the nature of scientific research, lessons that can help us all make more informed decisions.

A role for philosophy of science

Scientists are expert in the discipline they have been trained in. Their training and education introduces them to a wide range of techniques, methods, concepts, problems, solutions and problem-solving strategies. Philosophers of science may spend some of their time thinking about the same concepts, methods and so on, but philosophy of science is more centrally occupied with more general questions concerning what science is, what it achieves and how science differs from other ways of exploring the world. As we'll see throughout the book, when it comes to drawing conclusions about the state of a particular scientific issue, many of our most common mistakes stem from ignorance of how science works rather than ignorance of the science itself. For the purposes of utilizing scientific information responsibly,

unpersuaded that cigarettes cause lung cancer will be more likely to smoke. I leave it to readers to ponder on which explanation likely applied most broadly.

therefore, philosophers have accumulated valuable insights – insights that deserve attention when the credibility of certain scientific claims is up for discussion, insights that hold the potential to improve our capacity to evaluate what can appear *controversial* scientific claims and, thereby, our ability to successfully navigate through a complicated world. Thinking about what science is need not be a purely academic affair; a better feel for the nature of science holds the potential to change lives.

To help motivate the idea that philosophy of science might be relevant, let's consider the innocent observation that information is important or, more carefully, that *reliable* information is important. We hope that the accused are convicted on the basis of reliable evidence, that medical treatments are prescribed because there are good reasons to suppose they will be corrective, that social reforms and policies are introduced because there are indications that these will improve the circumstances of those affected and so on. Almost certainly we recognize that these hopes are not always realized (and perhaps we fear they're seldom achieved), but they certainly seem like the kinds of goals that we should be working towards. In all these cases, we require reliable information to reach reasonable decisions, and when we're in need of reliable information, we often look to science to provide it.

Admittedly, scientific conclusions alone can't, and shouldn't, dictate the outcome of practical decisions. Financial and moral considerations are two factors that may weigh against following to the letter what the science alone seems to recommend. But even before the point of introducing moral and economic considerations is reached, we might be confronted with a different kind of problem. Sometimes the scientists don't seem to agree. We read conflicting reports: about climate change, genetically modified food and vaccine safety; about the evidentiary support for evolutionary biology, the Big Bang model of cosmology and the geological column; about the costs and risks associated with various drugs and diets, educational and social policies. Our ability to act in accordance with the most reliable information is stymied by the absence of agreement between those who we look to for explanation and understanding. If we are to reach an informed decision on how we should act or what we should believe, then the appearance of controversy can represent a serious obstacle.

Sometimes, of course, the appearance of disagreement within a community of experts reflects a genuine lack of consensus among them. We'll see in due course that data are often vulnerable to alternative interpretations; scientists might have different ideas concerning what methods are most appropriate for a given project, what analysis is most sensible and what conclusions can reasonably be drawn. Controversy, debate and discussion are integral to the scientific process. On some occasions, however, the *appearance* of scientific controversy may be profoundly misleading and potentially enormously damaging. How many smokers continued with their habit during the second half of the twentieth century, in part because, as we've seen, the tobacco industry marshalled a successful campaign to increase doubt surrounding the established risks with their product? If doubt is influencing our decisions, but those doubts are being *created* by groups that desire to shape our behaviour, then we are being manipulated – and in ways that could potentially cost us a great deal. Learning to distinguish genuine controversies from the mere illusion of a controversy becomes essential to a proper appraisal of the state of the science, and hence the appropriate use of available information.

Reliable information is important, but distinguishing reliable from unreliable information is not always easy, where sometimes our confusion is the product of efforts to purposefully obscure the truth. Distinguishing genuine scientific disagreement from its imitations can be hard work. Part of the problem is that we can't escape the conflicting reports, whether in the media, on the internet, or at the water cooler. We're aware of rumours, apparently confounding evidence and contradictory advice. Merely sweeping aside one set of opinions can feel churlish, dogmatic and intellectually irresponsible. We might feel some sense of obligation to engage both sides of the debate, or at least suspend judgement until we've found time to properly evaluate the merits of each. We might suspect that there couldn't even *be* the appearance of controversy unless there was at least some merit to both sets of opinions. Would the connection between autism and vaccinations receive all the attention it does if there wasn't at least something to it? (The answer is that, yes, it does receive lots of attention despite medical professionals having unearthed not even a shred of good evidence to suggest that vaccinations increase the risks of autism.)

How then should we respond to contradictory reports, in our pursuit of a reasonable evaluation of the issues at hand? We could just wait for everyone to reach a consensus before engaging in any particular course of action, but this is risky. Some debates are so protracted that waiting for all the experts to agree would be reckless. Many decisions must be made within a given time frame. If we are relying on experts to help us make decisions, then it may be necessary to choose in the absence of complete consensus. An alternative and natural response to apparent disagreements within a scientific community is to better educate ourselves on the relevant science and then draw our own conclusions. We live in the information age. Never before has it been easier to access scientific journals, technical reports, articles and reviews. Of course we need to distinguish reliable from unreliable sources, but that just requires more information, of a slightly different sort. Furthermore, one might plausibly suppose that much public confusion stems from our collective ignorance of the relevant science. Depressing polls reveal that our scientific acumen falls far below what many would deem acceptable. (Several surveys have each found that a far-from-trivial proportion of people in many developed countries believe that the Sun orbits the Earth!) Disappointing levels of ignorance and confusion plague the kinds of scientific issues that are more central to this book. Surely we just need to educate people better on the science, one might think. Sensible as this sounds, however, as a response to the appearance of scientific disagreement, it has significant limitations; principally, it simply isn't sensible to suppose we can become sufficiently informed to make independently plausible and reliable evaluations of complicated scientific issues.

As science has advanced it has become increasingly specialized, so much so that not only is a professional molecular biologist likely to be quite unschooled on the latest developments in solid state physics, but individuals often pursue research that is largely incomprehensible even to others who work in the same general discipline. The pace and magnitude of scientific advancement is formidable. Matters are sufficiently involved even when we restrict our attention to the kinds of issues that are central to this book. The most recent report from the Intergovernmental Panel on Climate Change was an enormous effort, with contributions from hundreds of authors, reviewers and editors – their expertise ranging over a wide variety of disciplines related to understanding recent climate change, its impacts and available responses. And climate change is just one (admittedly conspicuous) topic that churns up a storm of public confusion and uncertainty. Additional environmental issues, health issues, not to mention evolutionary biology, are all areas where public evaluations of the science are markedly different from the attitudes of informed experts.

The difficulties associated with becoming sufficiently expert across a wide range of scientific disciplines are further aggravated when we attend to the importance of *tacit* knowledge within the sciences. This is a concept that was described several decades ago and concerns our abilities to reach reliable conclusions, or complete certain tasks, via methods that are not always easily communicated. Most of us know how to ride a bike, for example, but this knowledge is hard to make explicit in verbal terms, such that those who don't already know how to ride a bike can acquire that knowledge. Becoming expert in a given scientific discipline likely involves not just becoming better informed, but also working through problems, hands-on experience and a certain degree of immersion in the field. Learning to hit baseballs that are moving unpredictably and at high velocities isn't achieved solely by reading books. Similarly, learning to evaluate technical, scientific arguments, with respect to their methods, analysis, assumptions and conclusions, also requires experience and practice, and the development of skills that will not always transfer well from one scientific discipline to another. The required investments of time and effort associated with becoming adequately informed and rehearsed on subjects ranging from toxicology to palaeontology to atmospheric chemistry to epidemiology to molecular genetics to immunology to geophysical fluid dynamics to biogeography to virology to animal behaviour to hydrometeorology, and so on and so on and so on are surely impossible for any one individual.

Thus while craving reliable information in the face of controversy is entirely appropriate, neither waiting for all the experts to agree, nor rectifying all deficiencies in our own understanding of the science look like promising strategies for overcoming public confusion. This might seem like cause for dismay. However, much of the troubling confusion we'll be concerned with in this book is a product of wishful thinking and other cognitive biases, poor critical thinking and general ignorance (or convenient memory loss) of how science (in general terms) actually works. The principal theme of the book is that we would all respond to the appearance of controversy more appropriately, avoid error more reliably and thereby keep ourselves, our families, our communities and our planet more healthy if we better understood the nature of scientific inquiry, how easy it is to distort its findings and how some special interest groups have made an art form of deceiving us. A better grasp of the relevant science certainly wouldn't hurt; a better sense for the nature of science, in conjunction with improved critical thinking skills and greater awareness for our own cognitive shortcomings, would help more.

Created controversies

By the mid-1950s there existed no substantive controversy surrounding the connection between cigarettes and lung cancer within the relevant scientific community. However, tobacco companies managed to create, or manufacture, the *appearance* of scientific controversy. Part III of the book provides extended discussions of several further, putative, examples of *created controversies*. Much of what precedes will serve as groundwork. In Part I, we'll survey important themes in philosophy of science, and thereby identify common misconceptions about the nature of scientific inquiry and start developing a more sophisticated understanding of how science works. A better appreciation of the nature of scientific thinking helps with assessments of manufactured controversies, but improved critical thinking more generally is also important. One way of motivating the importance of stronger reasoning skills are some truly absorbing conclusions that have emerged from work in cognitive and social psychology. To illustrate, consider the following vignette, presented to subjects in the 1980s by psychologists Amos Tversky and Daniel Kahneman:

Linda is 31 years old, single, outspoken, and very bright. She majored in philosophy. As a student, she was deeply concerned with issues of discrimination and social justice, and also participated in anti-nuclear demonstrations.

Subjects were asked to evaluate which seemed the most probable:

A) Linda is a bank teller
B) Linda is a bank teller and is active in the feminist movement

A strong majority responded that the second option was more probable. They were wrong. To see why, notice that if we attend to all those possibilities in which Linda becomes a bank teller, then we are certain to *include* all those possibilities when Linda *both* becomes a bank teller and enters the feminist movement. Linda can't become both a bank teller and active in the feminist movement without becoming a bank teller.

Thus there are more ways of becoming a bank teller than there are ways of becoming both a bank teller and active in the feminist movement. Consequently, it must be more probable that Linda just becomes a bank teller.

The particular error in reasoning that Tversky and Kahneman discovered has been investigated in subsequent studies. It has become known as the conjunction fallacy. The result is remarkably robust. In further tests a significant proportion of physicians and medical students committed the fallacy. In another, graduate students from the University of California, Berkeley, and Stanford University, all of whom had taken several statistics courses, were asked the same questions about Linda. Over a third got the answer wrong. We are all vulnerable to committing the conjunction fallacy, in which we conclude that the probability of multiple events is higher than the probability of one of the included events. Tversky and Kahneman were pioneers in the study of systematic human cognitive biases. Dozens of such biases have now been uncovered, some of which are quite relevant for purposes of understanding why our ability to reliably evaluate scientific claims is often compromised. We will consider some of these in more detail in Chapter 5. One of the general lessons to draw from such studies is that, in at least some circumstances, we should be extremely wary of our own intuitions, evaluations, assumptions and judgements. We must learn to mistrust our judgements, in certain kinds of circumstances, and seek more reliable means of reaching sensible conclusions. In short, we should desire better evidence and better arguments than personal experience and deliberation can provide.

In Chapter 6 we'll consider the form of arguments, the way arguments can go wrong, some of the classical informal fallacies that we should be alert to and some basic strategies we can utilize for purposes of better evaluating arguments and debates. Chapter 7 provides a more detailed analysis of the concept of a created controversy. By the end of the second part of the book we should have achieved a greater appreciation for how science works and what it can and can't accomplish, how the appearance of controversy can mislead, why we should be cautious about our intuitive reactions to certain arguments and ideas, some of the most common errors of reasoning that lead people astray and some basic strategies for assessing claims and evidence. Furnished with such information and skills, we will be better situated to launch into considering cases that are regarded by many as paradigmatic, created controversies.

The primary purpose of this book is thus not to persuade readers that anthropogenic climate change is real, nor to provide a definitive evaluation of the merits of Intelligent Design theory. The book is not designed to convince readers that cigarettes are dangerous, that abstinence-only sex education programs don't work, that HIV causes AIDS or that autism has nothing to do with vaccines. There are already excellent books and articles on these issues, which are accessible, thorough and written by experts far better qualified than I am to write on such topics. Nevertheless, although these books and resources have enormous value, they also have limitations. For example, negative responses to such books on blogs, in reviews, by word of mouth and so on draw attention to some new study that has occurred since the book was written, or some (apparently) important articles that the author didn't engage with. Such studies and articles, it is then suggested, stand in direct opposition to the conclusions of the book, undermining those conclusions and the evidentiary merits of that particular cause. The correct response to such reviews, however, isn't always an updated edition of the book, complete with answers to each and every objection. What's needed, as noted earlier, is for us to respond better to the appearance of doubt. Furthermore, the illusion of controversy will appear elsewhere, perhaps courtesy of some cohort that hopes to exploit our doubts and vulnerabilities. Being forewarned about such possibilities will help us respond more sensibly. Rather than argue strongly for any particular views across a range of issues, the objective of this book is instead to encourage and cultivate critical, honest and reflective thinking and, where necessary, openness to revising prior attitudes.

Conclusions

An alternative perspective on the general topic of the book can be achieved by reflecting on two attitudes towards science that are blatantly inane. The first involves an unabashed, uncritical and unfiltered endorsement of all and any conclusions that are announced in the name of science. Such reverence is undesirable, by science's own standards, and leads quickly to inconsistency. Science self-describes as cautious, self-critical and self-correcting. All aspects of the scientific process are, at least in principle, open to criticism and revision. We'll have cause to consider whether scientists are as willing to revise their attitudes as we think they should be, but, even if we discover more dogmatism than we might have expected, we won't find anyone uncritically accepting *every* putative scientific claim. The primary reason, of course, is that scientists don't always agree. Anyone inclined to impulsively endorse all scientific conclusions will soon become dizzy with contradictions.

Carelessly embracing all scientific proclamations is undesirable. At the other extreme is an equally blanketed rejection of scientific claims, an attitude that might sound just as objectionable and similarly problematic. Such radical scepticism has been suggested and has even seemed plausible to some. Perhaps a more common attitude, less radical in the breadth of conclusions it rejects but no less worrisome, seeks to discriminately challenge only those scientific conclusions that clash with agendas that are quite unrelated to the pursuit of truth and understanding. If a corporation desires to maximize profits but is confronted with evidence that their product is harmful, then its board members might dogmatically challenge the science only *because* it threatens profits. The health of its customers is collateral damage.

Neither uncritical acceptance nor unjustified rejection of scientific conclusions is helpful. It is also potentially dangerous. A middle path between these extremes is needed but, for a variety of reasons, not easily found. A particular confounding factor that is central to this book concerns those occasions when the appearance of controversy is unrepresentative of the relevant scientific community's attitude. Ignoring our experts, and attaching undue weight to the opinions of those who have independent motivation to promote a particular set of ideas, is irresponsible and risky. We must work hard to expose created controversies for the illusions that they are.

Discussion questions

1. How much does the layperson understand of why cigarettes cause lung cancer? Think carefully about your own level of understanding. What might this suggest about how facts become accepted within society?
2. How much science is it reasonable to expect people to learn, understand and retain? How does this relate to the level of trust we should place in scientific authorities?
3. Why do you suppose so many people were unconvinced of the connection between cigarettes and lung cancer for so long? With hindsight what, if anything, could have been done differently to prevent the widespread confusion?
4. Why do we need a theory of science, that is, an account of what science is, what it achieves, what it assumes and so on?
5. Are genuine (rather than created) scientific controversies a cause for concern? How should we respond to compelling evidence that a genuine scientific controversy exists?
6. Some laypersons find the consequences of modern physics so peculiar that they conclude the theories must be flawed. Are there differences between this example and Galileo's contemporaries rejecting heliocentricism? Are there cases where it would be appropriate for a layperson to reject scientific theories based on how outlandish they appear?
7. Why do you suppose there exists considerable public scepticism surrounding some scientific claims, but very little surrounding many others?
8. Does exposing created scientific controversy matter only if it has clear, undesirable, practical implications?

Part I

Lessons from the Philosophy of Science

1

Defining science and the empiricist approach

Insofar as cigarette companies funded research principally to protect industry interests, rather than launching a sincere effort to properly evaluate the health risks associated with their product, we might come to regard their behaviour as unscientific. It is a common feature within discussions of Creation science, evolutionary biology, environmental and health sciences, the search for extra-terrestrial life, ghosts, Bigfoot and so on, that some or other practice, theory, method or conclusion is admonished for mimicking science but falling short of scientific ideals. The prestige that modern society affords science practically guarantees that impassioned debates will erupt over whether some discipline or other does or does not belong. Such sentiments might suggest that the most important task in this book should be that of defining what exactly science is.

Furthermore, it feels quite natural to suppose that, given a little time and ingenuity, we should be able to define science, and thereby distinguish it from other ways of knowing and understanding the world. We talk about science as if the word picks out a single, unified subject, method or discipline. Our schools and universities are organized as if more specialized fields and subject matter either unambiguously do or do not belong to the sciences. We describe a history that appears to enforce the same, general attitude. That history typically begins with Nicolaus Copernicus who, in 1543, published *On the Revolutions of the Heavenly Spheres*, in which he presented his heliocentric model of the solar system. Although it took almost a century before heliocentricism was widely accepted among scholars, that date is often identified as the start of the Scientific Revolution and the birth of modern science. The phrase itself – 'the birth of modern science' – implies that something new appeared, something that has subsequently grown and developed, but, again, some *thing* that we might hope to identify and define.[1]

[1] Historians of science now offer a much more nuanced account of the changes and continuities that surround the origin, reception and development of heliocentricism and the revolution it supposedly induced. Historians are particularly sensitive to the possibility that hindsight has an important, but misleading, influence over our reflections on the history of ideas and our judgements concerning whether certain activities were scientific or not. For a wonderful introduction to the history of science, see Bowler and Morus (2005).

One source of trouble for our project arises, however, when we notice that defining science in a historical, or institutional, sense is quite distinct from the project of understanding science as our most reliable means of acquiring knowledge. To illustrate, suppose that we find a reasonable definition for what Copernicus, Galileo, Newton and their intellectual descendants have been doing. We announce that science requires fidelity to these particular methods and assumptions. Suppose now, however, we are presented with studies that don't fit the definition. One reaction would be to rule that this absence of fit disqualifies such studies from scientific status and thus renders them undeserving of further attention or scrutiny. But that ruling seems hasty. Perhaps there's a problem with our definition, and even if we're convinced there isn't, if the studies appear valuable, sensible and capable of producing genuine insight, then surely they deserve further analysis regardless of their relationship to our definition. To think otherwise doesn't seem *scientific*.

We've uncovered an apparent tension. In one sense, research programs that fall outside our definitions of *science* don't qualify as science. However, dismissing these programs too hastily seems unscientific. Early twentieth-century medical science didn't include statistics, but we admire those who were willing to consider the merits of the new approach over those who stubbornly refused. Seemingly, behaving scientifically involves judging ideas according to their merits, but this would appear to involve more effort than merely confirming whether or not those ideas satisfy a given definition.

The word *science* is used both to describe a set of practices and methods, but, as we've alluded to already, it also carries a strong normative component. Providing *scientific* evidence for some conclusion has rhetorical force. Achieving scientific status is important to nascent disciplines and fringe ideas. Efforts to better understand the nature of science must therefore straddle both descriptive and normative concerns. We want to better understand the nature of science, but also what justifies scientific authority. We can attempt to describe science, in terms of pedigree, history, shared commitments and practices or even in terms of institutional structure, influence and power. However, our descriptions won't provide any immediate reasons to *prefer* scientific conclusions. If we are interested to defend scientific authority, then our discussion must extend beyond descriptions and offer some account of why scientific conclusions are superior. The relationship between descriptive and normative philosophy of science complicates our hopes of distinguishing science from alternative ways of knowing. Nevertheless, it is to this project that we now turn.

The problem of demarcation

The problem of demarcation permits of several interpretations, but the overarching concern is to differentiate science from other ways of generating and defending beliefs. Some philosophers and scientists have been motivated to distinguish science from all non-science; others have had a narrower focus and attempt to delineate science from religion, or science from metaphysics, or science from pseudoscience. The problem's central role within philosophy of science is attributable to its promise to differentiate reliable from unreliable beliefs. Science, let's assume, deserves its privileged status as our most dependable means of acquiring knowledge and understanding. If, therefore, we could identify those features of science that distinguish it from non-scientific investigations and methods, a lot of debates could be resolved.

By locating the proper *definition* of science, we can determine whether particular studies, theories and results can be justifiably labelled *scientific*. If in applying our definition to particular disputes we discover that the arguments and theories advanced on one side of the debate are unscientific, while the opposing side can legitimately lay claim to scientific support for their conclusions, then this would appear to at least help settle the dispute. The parents of some autistic children have suggested that vaccines caused the autistic disorder. If the putative connection between vaccines and autism is not based on scientific evidence, however, but refutations of that connection are, then these judgements – about what is or is not science – might in themselves be offered as evidence that the vaccine-autism connection is little more than scare-mongering. Likewise for the benefits of acupuncture, the risks of too much fluoride in drinking water, the reliability of expert testimony in legal cases, the evidence that supports the Big Bang theory and the credibility of astrology, phrenology and homeopathy.

Solving the problem of demarcation might thus have enormous practical import, because not only could a solution help settle issues of great social significance, it might do so in ways that should be comprehensible even to those who lack significant scientific training. We need only provide people with the proper definition of science, so they can all see whether something lacks scientific standing, and thus whether its central claims should be ignored, or at least regarded with greater suspicion. Of course, the benefits of answering the problem aren't a good reason to suppose that a solution is possible. In fact a compelling answer has remained elusive, and there are reasons to wonder whether we should in fact care to resolve the problem at all. It is worth reviewing some proposed solutions before returning to reflect again on our motives.

One feature of science that appears distinctive is scientists' tendency to perform experiments. Astronomy is clearly a thoroughly scientific pursuit, however, but it is rarely involved in the kind of experiment that involves manipulating causal factors, managing controls or randomizing across groups. Similarly, field scientists gather careful, thorough observations of organisms in their natural habitat for example, but again rarely conduct controlled experiments. Many scientific disciplines have an important historical dimension. We're interested in why the dinosaurs became extinct, and how the Earth's atmosphere, the Grand Canyon and the moon were formed. Certain experiments might teach us a great deal about these questions, but often our best answers will draw on a wide variety of methods. We could easily place too much emphasis on the role of experiments in science, and consequently produce an impoverished view of what science is.

In response to these apparent counterexamples, we could adopt a more inclusive idea of what experiments involve, such that the principal methods employed by astronomers and field scientists would now qualify as experiments. But this strategy runs the risk of making almost any effort to acquire information into an experiment. We may talk, informally, about a chef experimenting with new ingredients, or an artist experimenting with different techniques, but neither of these activities are regarded as scientific. On any definition of *experiment*, it seems likely that there will be either examples of science that doesn't involve experiment, or examples of non-science that does, or maybe even both. The success of any proposed solution to the problem of demarcation will be evaluated, in part, by how closely it aligns with prior judgements about what does and does not qualify as science.

The most famous effort to demarcate science from non-science was described and defended by Sir Karl Popper. Popper believed that science is centrally concerned with *falsifying* scientific theories, rather than gathering evidence to support them. He offered falsifiability as a novel means of understanding the scientific method and as a criterion for delineating genuine science from other ways of investigating. Neither suggestion has subsequently been regarded as successful, at least among philosophers, but it is his demarcation criterion that we'll focus on here.

A falsifiable theory is one that we could discover is false by making observations that are incompatible with the theory. Falsifiability was the essential property that genuine scientific theories possessed and non-scientific theories lacked, according to Popper. He illustrated the idea by contrasting a scientific theory that he thought laudable (Einstein's theory of General Relativity) with ones for which he held a much poorer opinion (Freudian psychoanalysis, for example). An important prediction of Einstein's theory is that light is deflected within the gravitational fields of massive objects such as the Sun. Typically this effect is unobservable, but during a solar eclipse it is possible to test the prediction. In 1919, Sir Arthur Eddington, a British physicist, led an expedition to observe such an eclipse and walked away claiming a huge triumph for Einstein and his theory. In Popper's opinion Einstein's success was a product of his bold and precise prediction about light bending. It was Popper's conviction that, had Einstein's prediction not been verified, the scientific community would quickly have lost interest in General Relativity.

Popper's complaint against pursuits like astrology, phrenology and others is that they don't make 'bold predictions', only vague general claims about the world that can be reconciled with any subsequent body of evidence. According to Popper, Freud's theory was consistent with all human behaviour. If a man pushes a child into a lake with the intention of killing the child, Freudians can explain the behaviour in terms of their concept of *repression*. By contrast, a man who jumps into a lake to save a drowning child, knowing that he will likely sacrifice his own life in the process, has *sublimated* unconscious impulses. However someone behaved, opined Popper, Freud's theory could explain the behaviour in terms of unconscious desires. Freudian theories thereby appear unfalsifiable. Einstein's theory wasn't falsified by Eddington's observations, and Freud's theory wasn't falsified by observations of human behaviour. The important difference, for Popper, is that Einstein's theory *would* have been falsified if Eddington had failed to detect the predicted degree of light bending, whereas Freud can't be falsified no matter how people are observed to behave. Popper justified his own views about which theories deserved admiration, and which deserved contempt, by arguing that the former are falsifiable while the latter are not.

Despite some attractions, the falsifiability criterion faces critical objections. The falsification of simple, universal statements might appear relatively straightforward. The hypothesis *All ravens are black*, for example, is seemingly falsified by a single observation of a white raven (or a pink, yellow, blue or mauve one). However, the falsification of a mature scientific theory, against the background of assumptions that are involved in testing that theory, is rarely so unequivocal. Confronted with unexpected, anomalous, potentially falsifying data, there are always ways of rescuing a troubled theory from falsification, by shifting the blame to other beliefs and commitments.[2] If we relied on delicate apparatus for purposes of gathering our data, for example, we may suspect that the apparatus likely malfunctioned, or that its reliability was compromised by external shocks. Our experimental design might include theoretical assumptions that are mistaken. Even our naked-eye observations can't always be trusted, a worry we will return to in the next chapter. In short, inconsistencies between our data and the predictions of a scientific theory might arise for many reasons, so a given theory or hypothesis can only be conclusively falsified when we've ruled out all other possible sources of error. But ruling out *all* possible sources of error is clearly an impossible standard.

Not only is it possible to rescue theories from falsification, sometimes such liberations appear entirely reasonable, and can even lead to important scientific achievements. In the mid-nineteenth century the orbit of Uranus was observed to deviate from predictions based on Newtonian physics. One suggestion tendered was that perhaps an additional, massive and as-yet-unobserved planet existed beyond the orbit of Uranus. The gravitational tug of the hypothetical planet could conceivably account for the perturbations in Uranus's orbit. On the basis of Newtonian physics and the observed orbit of Uranus a new planet, Neptune, was discovered, a striking achievement for the physics of the day. The unexpected observations concerning the orbit of Uranus could have been regarded as a falsification of Newtonian physics, but such a reaction looks, particularly with hindsight, hasty.

[2] This problem for Popper originates in what's known as the Duhem-Quine thesis, which we will return to in Chapter 4.

A more recent example involves the CERN researchers who, in 2011, claimed evidence of particles travelling faster than the speed of light, a noteworthy report because superluminal particles are inconsistent with expectations based on our best theoretical physics. One reaction to the announcement would've been to immediately reject the theories, stop teaching them in universities and shred the textbooks that describe them. No-one advocated such a reaction. A common response was to doubt the data, a response that was vindicated a few months later when specific problems with the research were discovered.

These examples illustrate how giving up on a scientific theory too quickly might be just as foolhardy as persevering with a theory in the face of overwhelming evidence that the theory is flawed. Perhaps there's some comfort in supposing that any given theory could, in principle, be decisively falsified by a single crucial experiment, but there are compelling reasons to doubt that science does, or should, work that way. This is not to say, of course, that theories aren't sometimes falsified and replaced, but anticipating ahead of time exactly what will induce the rejection of some theory seems unlikely. However, if we can't make predictions about what possible observations would falsify a scientific hypothesis or theory, then we can't use falsifiability as an effective demarcation criterion. Almost all ideas are falsifiable if confronted with *some* appropriate collection of confounding evidence. If falsifying *scientific* ideas is less ambiguous than the falsification of pseudoscientific or non-scientific ones, then no-one has made that argument convincingly.

The difficulties with Popper's falsification criterion might suggest that what distinguishes science is not falsifiability but *testability*. We can continue testing a theory, after all, even if we're unsure ahead of time what battery of tests might falsify it. Testing also seems central to scientific investigation. As a demarcation criterion, however, testability is also problematic. Many paradigmatic instances of pseudoscience are testable. Some Creation scientists, for example, assert that the Earth is just a few thousand years old, a belief that can be tested. Those who drew a connection between thiomersal (a preservative that is widely used in vaccinations) and autism predicted a decrease in the number of new cases of autism in California following that state's decision to ban vaccines that contained it. The prediction was not borne out.

Almost all experts were, and remain, critical of the putative connection between thiomersal and autism. However, insofar as some may have regarded the idea as unscientific, that judgement was clearly not based on a failure of the hypothesis to yield testable predictions. With a little ingenuity, scientists have found ways of testing even those belief systems that have the reputation of offering only hopelessly vague and untestable predictions. For example, astrologers have been challenged to match personality profiles with astrological charts. They performed no better than would've been expected on the basis of chance alone. Astrology failed that particular test, but only because we found a way to test it.

As Larry Laudan astutely observed, hypotheses that fail every experiment we design to test them must be testable.[3] An important lesson lurks within this seemingly trivial observation. Solutions to the problem of demarcation are supposed to help us distinguish science from alternative ways of knowing, but this won't always help justify the authority we assign to scientific conclusions. Even if testability had been able to help us distinguish scientific ideas from non-scientific ones, it does very little to recommend the former. The same problem plagues Popper's criterion: even ignoring the above problems, it is unclear whether a theory's being falsifiable is *itself* relevant to evaluating a theory's merits. We admire theories that pass many, and stringent, tests. Likewise, we will be interested to learn that a theory has been falsified.[4] We can't test the untestable or falsify what's unfalsifiable, but we must be careful to distinguish the qualities we judge relevant to our evaluations from the preconditions of making those evaluations. We will return to this worry below.

Given our ambition to identify some essential feature of scientific inquiry we should reflect on the extremely pervasive idea that what makes science distinctive is its fidelity to *the scientific method*. It is often repeated that scientists employ the scientific method, but as a putative solution to the problem of demarcation this appears just to *rename* the problem, or beg the question. That's to say, anyone who seeks to better understand what separates science from non-science should be dissatisfied with appeals to the scientific method, until it has been made clear exactly what that method amounts to. Second, standard accounts of the scientific method typically appeal to the same kinds of considerations that we have already discussed: testability and falsifiability, for example. Efforts to better understand the scientific method thus run into the same problems we've already encountered. Third, as we saw with the research surrounding the health risks associated with cigarettes, the methods accepted by a given scientific community can change over time. Scientific methods are themselves an appropriate object for evaluation, critique, amendment and improvement. Perhaps if we retreat to sufficiently general concepts, we will see that all scientists employ the same method, regardless of their goals, discipline and training, but it might be more useful to think of scientists not as employing a single, unified *scientific method*, but as raising questions and then seeking the most suitable methods they can find, borrow or develop for purposes of making progress in answering those questions. By entertaining this proposal we start to move away from thinking of distinct scientific disciplines as all possessing a unique common feature.

We've reviewed a number of candidate solutions to the problem of demarcation. None provide straightforward and reliable means of distinguishing science from non-

[3] Laudan (1983).

[4] A further important objection to Popper's more general philosophy of science is that he lacks any convincing resources for comparing alternative theories that have not been falsified. Suppose one hypothesis has passed many tests, but an alternative has passed none. We are going to regard the former more favourably, but all Popper seems able to say is that neither has been falsified, leaving no room to justify our preference. Popper did attempt a response to the objection, introducing a notion of 'corroboration', but few have regarded the response as successful.

science. Testability and falsifiability are ineffective demarcation criteria. Appeals to the scientific method are too vague to offer much help. The long history of failed attempts to solve the problem has convinced some philosophers that there is no solution and that the problem is no longer worth engaging. Importantly, these philosophers do not claim that we can't distinguish plausible conclusions from implausible ones, only that attempts to identify the defining feature, or small set of features, of scientific inquiry aren't helpful means of accomplishing this goal. The history of failed attempts isn't the only reason to suspect that a solution to the problem might be unattainable.

Historical and sociological evidence suggests that science is too heterogeneous for any simple definition to divide all that we recognize as science from that which we don't grant scientific status. Many distinctions have been drawn between ostensibly different types of science: natural versus social, experimental versus historical, soft versus hard, theoretical versus experimental and so on. That so many thinkers have thought it useful to develop such distinctions might again seem to imply that attempts to force all and only science into a single definition might be unhelpful. Finally, we've alluded several times to the fact that science changes over time, not just in terms of the theories it endorses but also of the methods it employs and standards it adopts. The early twentieth century saw growth in the use of statistical methods within many scientific disciplines. Over the last few decades the use of computer modelling has seen a similar upsurge. If our definition of science is prone to change, or our stock of acceptable scientific methods is subject to revision, then we can't place too much emphasis on any definition that seems apt at the moment. New scientific techniques might force revisions to our understanding of how reliable conclusions can be generated, contravening the idea that we can justifiably ignore any practice purely on the grounds that it doesn't fit our current definition.

Nevertheless, although many philosophers of science seem to agree with Larry Laudan's announcement that the problem of demarcation is in demise, there are noteworthy exceptions.[5] John Dupré, for example, agrees that science is not a unified set of methods and practices, and he acknowledges that this excludes the possibility of a simple criterion of demarcation.[6] However, he rightly insists that scientific authority should be defended, and suggests that we might achieve this by understanding science as a cluster concept. For Dupré, the virtues of science include its 'sensitivity to empirical fact, plausible background assumptions, coherence with other things we know, exposure to criticism from the widest variety of sources, and no doubt others. Some of the things we call *science* have many such virtues, others have very few.'[7] On

[5] Laudan (1983).

[6] There is far more interesting conversation surrounding the problem of demarcation than it is sensible to review here. I am generally sympathetic to Laudan's conclusion, but see Pigliucci and Boudry (2013) for a range of alternate perspectives.

[7] Dupré (1995, 243).

Dupré's view there might be no feature that all sciences share, nor a feature that non-sciences all lack, but if we attend to the right set of features then we'll find that the sciences have more of them, or instantiate them more clearly. Dupré's suggestion is important, but rather than explore it further, I want now to return to our original motivation for concerning ourselves with the problem of demarcation. Ultimately, I'll suggest that whether we understand ourselves as abandoning the problem of demarcation or analysing science as a cluster concept doesn't actually matter a great deal.

Our ambition is to hold reasonable beliefs. We want to avoid falling for the rhetorical skills of charlatans, quacks and snake oil salesmen, as well as the claims of those who might be entirely well intentioned but who continue to perpetuate ideas that are both false and dangerous. By limiting ourselves to *scientific* beliefs we hoped we could achieve this ambition, at least to a significant extent, but the strategy requires clear criteria that most science satisfies and most non-science does not. Describing key differences between science and non-science is not, however, quite sufficient. First, if our solution to the problem of demarcation is to facilitate better reasoning within public discourse, enable laypersons to make better decisions, recognize pseudoscience for the deception that it is and so on, then these criteria must be widely comprehensible and accessible. If the criteria by which we distinguish science from its imitators become sufficiently cumbersome, then we might wonder whether the layperson would be better served by an introduction to the relevant science, rather than a lengthy explanation about why some or other set of ideas lacks scientific status. Second, it is reasonable to seek not just criteria that partition science from non-science in ways that agree, at least reasonably well, with prior attitudes. In addition, we may desire good reasons to believe that the presence or absence of those criteria is relevant to the theory's plausibility.

The problem of demarcation involves an uneasy mix of descriptive and normative objectives. At the poles are perspectives that illustrate the difficulty. If we attempt simply to *describe* the most salient differences between science and the alternatives, we have not thereby offered any argument for scientific authority. If we seek instead a purely normative theory of science, then we may find that we lack the resources for assessing the kinds of issues that matter to us. Testability is perhaps an epistemic virtue, but it is so frequently satisfied that it does almost nothing to help us reach sensible decisions. There are different ways in which hypotheses can be tested, where some tests are more telling than others. There are different ways to gather data and important differences in the quality of data. There are different sources of evidence and distinct mechanisms by which these can help support hypotheses. Once we start attending to such details, we are no longer seeking a few, readily identifiable features of science. Yet these are the kinds of details that often are extremely pertinent.

Collectively we cannot ignore the many ways in which scientific arguments, methods, analyses and practices have been developed, improved and refined. We are soon forced to recognize that theories can fail or succeed for a wide range of reasons. Perhaps it was once reasonable to hope that a simple demarcation criterion would have sorted science from non-science unambiguously, while simultaneously providing a reasonable justification for our greater confidence in the former, but that ambition no longer appears sensible. We can start cataloguing the kinds of mistakes that are most common, the methods that are most widely used, the reasons some methods have been modified and so on. But as justifications for science become more comprehensive, and more specific to particular disciplines, so it seems to matter less whether we label something as non-science, bad science, pseudoscience, nonsense or gibberish. This isn't to deny that there might be important distinctions to be made here, and lessons worth learning. However, determining which *labels* are most appropriate is likely to be less helpful than reflecting on the kinds of arguments and evidence that are made within the context of the kinds of debates that concern us. The project of defining science cedes to the project of identifying which beliefs about the world are plausible in light of available evidence.

That no simple solution to the problem is available is cause for disappointment, excitement and surprise. Disappointment, because life would be easier if unreliable results and methods wore their sins on their sleeves. It would be hugely convenient if we could distinguish scientific from unscientific conclusions at a glance, and concurrently defend the idea that scientific conclusions are more trustworthy, for we could then bypass a lot of complications. The excitement occurs because we are now required (collectively) to listen carefully to all ideas, suggestions, proposals and conjectures, evaluate the arguments and evidence that are advanced both for and against and assess them directly. We won't rest content with dismissing novel methods simply because they don't fit our definitions. Individually, we are too puny to achieve such an ambitious undertaking. We must learn how to better utilize the information we can reasonably keep abreast of. The absence of a solution to the problem of demarcation is, furthermore, highly surprising. We're so accustomed to talking about science as if it's some monolithic entity that we are inclined to suppose that a solution to the problem must be possible. Only by thoroughly embracing the possibility that science might be better conceived as a dynamic *collection* of theories and assumptions, more or less loosely connected, employing different methods, in pursuit of different goals, and always prone to revisions that we can anticipate in neither detail nor scope, are we likely to overcome this potentially misleading tendency. To try to help sell this adjustment in thinking, it might be sensible to desist from talking about *science* and refer instead to *sciences*.

One further facet of the problem of demarcation is instructive. In discussing the falsifiability criterion we saw that for a theory to be falsified it must be falsifiable. On certain issues, however, it seems the problem we face is not that certain theories are unfalsifiable, it's that *advocates* for those theories refuse to admit what is obvious to everyone else – that the theory has been falsified. Similarly, sometimes the problem is not that some are willing to make claims that scientists can't test, it's that scientists' conclusions are ignored, or distorted, or suppressed, on those occasions when the conclusions don't fit an independent agenda. The implausibility of certain claims often isn't immediately apparent to the layperson, but the dogmatism with which these are asserted by some, in the face of careful, compelling objections from experts in the field, suggests an inability to evaluate the issues rationally. Proponents for certain ideas are unresponsive to objections, and obscure the nature of their main claims.

All these problems, however, principally concern the characters of those individuals who defend the ideas, rather than the ideas themselves. It of course does little to recommend a theory if its advocates are irrational, ignorant or dogmatic, but the curiosity of bystanders to these debates won't be satisfied by such name-calling. If it is irrational to deny climate change, or the cigarette-lung cancer connection, then we will want to know why. Satisfying such curiosity will involve at least some engagement with the available evidence and arguments. Whether something fits a particular definition or conception of *science*, or whether the proponents of a theory are behaving reasonably and rationally, are each less important to us than whether the claims are plausible.

The empiricist attitude

Perhaps we can't define the sciences sufficiently precisely to settle the kinds of issues that concern us most, but we might nonetheless hope to improve our decision-making processes by reflecting further on the nature of scientific investigations. In the remainder of the chapter, we'll introduce an attitude towards the sciences that has very broad appeal, an attitude that places a strong emphasis on the importance and centrality within the sciences of *observations*, or *data*, and assumes that by combining reliable data with appropriate forms of reasoning, sciences achieve substantive and reliable knowledge. This basic empiricist attitude is a useful platform for understanding and developing more sophisticated theories of how the sciences work. An illustration is probably a good place to start.

In the 1850s, Gregor Mendel, an Austrian monk, living in Brno, began cross-breeding pea plants – lots of them. He'd selected this particular plant because of its discontinuous traits. Some of his pea plants had long stems, others were short, but none were intermediate in length. Some produced wrinkled peas, while others produced smooth peas. Within the plants Mendel used he identified seven such discontinuous features. The experiment began with Mendel self-fertilizing the original stock for several generations, to ensure that they bred true, that tall plants produced seeds that themselves became tall plants, for example. The next stage required crossing plants that possessed opposing traits. Wrinkled-pea-producing plants were crossed with smooth-pea-producing plants, and purple-flowered plants were crossed with white-flowered plants. The results were less than astounding. For each pair of traits, one was lost in the first generation of hybrids. Each hybrid produced only smooth peas, or only purple flowers, or had long stems, and so on.

Fortunately Mendel hadn't had enough of the labour-intensive research and began self-fertilizing the hybrid generation, and with results that would achieve enormous historical significance, albeit a significance that went unappreciated for several decades. Traits that had been lost reappeared, less frequently than the opposing characteristic, and always with a ratio of approximately 3:1. Mendel could have rested content with an interesting discovery about hybridization. He didn't. Instead he proposed three *principles* of inheritance, designed collectively to account for the data he had carefully and painstakingly gathered. In light of the data he'd gathered, these principles of inheritance seemed plausible. His methods also provided a means of independently testing the principles he had proposed. If Mendel's principles were correct, then backcrossing the first generation plants with some of his original stock should produce offspring with both discontinuous traits in approximately even numbers. The data Mendel gathered in response to this test supported his predictions.

Mendel carefully gathered data which suggested to him certain hypotheses, which he could independently test against further data. Insofar as we trust Mendel to have properly gathered and recorded the data, and regard his reasoning as reasonable, we might feel warranted in concluding that his principles properly describe something important about inheritance mechanisms. There are a slew of questions we can raise in response to Mendel's work. In broadest strokes, we should be concerned about the reliability of Mendel's data, his assumptions, the reliability of his inferences and the significance of the conclusions he reached. These concerns are pursued by formulating more focused questions. Did Mendel do enough to protect against accidental cross-pollination? Was his record keeping accurate?[8] Were others able to replicate Mendel's results?[9] Mendel had found a way of testing his principles, but how much confidence should we place in those hypotheses once we learn that they passed the independent tests? Had Mendel discovered something important about inheritance generally, or had he discovered something interesting only about inheritance mechanisms within plants, or within a particular family, genus or species of plant life? Mendel purposefully sought out discontinuous traits that happened to sort independently. To what extent did restricting the studies in these ways undermine the value of his conclusions? Are discontinuous traits common within nature? Could Mendel's principles be utilized to understand continuous traits?

[8] A famous, and widely discussed, analysis of Mendel's argument was offered by the Cambridge statistician, Sir Ronald Fisher, who argued, in the 1930s, that some of Mendel's data were 'too good to be true'. Fisher did not suppose that Mendel was guilty of deliberate fraud, nor that his criticism of Mendel's work detracted from the merits of either Mendel's conclusions or argument. For more on this interesting episode, see Franklin et al. (2008).

[9] As a result of corresponding with botanist Karl Nägeli, Mendel was himself convinced to try replicating his results, but, as it turned out, with a plant species that was not at all well suited for Mendel's pursuit. See Henig (2001) for more on this and other aspects of Mendel's life and work.

The development and exploration of Mendel's principles launched the field of genetics. Addressing some of the concerns led to the modification of those same principles. The history surrounding Mendel's work, its reception and its rediscovery is far more interesting than the sketch I've provided here. Hopefully, however, stripping the story down will help illustrate an attitude towards the sciences that is prevalent, appealing and influential. In addition, remembering that the story is diluted should remind us to be cautious with the morals we might wish to draw from such narratives.

Mendel's story illustrates the idea that what's central to science is the gathering of observations, sometimes under very carefully controlled conditions. Exacting measurements are sought. Reliable repetition of those results might be required. Such efforts might rely on new methods, technologies and sophisticated instruments, perhaps designed specifically for the task at hand. The result of these labours is a body of data that serve as a valuable platform for proposing hypotheses about the world. Once articulated, these hypotheses are subjected to rigorous testing and evaluation, part of which will require gathering more data. Those hypotheses that withstand close scrutiny achieve some higher level of credibility within the scientific community and society more generally. Measurements, observations and data thus *motivate* a line of scientific inquiry and subsequently allow us to *adjudicate* the plausibility of hypotheses and theories. If we accept those data as reliable representations of the phenomena we are investigating, furthermore, then theories that fit well with those data might strike us as good candidates for properly describing what's really going on. Observations thereby play a *foundational* role, *their* certainty and reliability providing the objectivity we desire for the theoretical structures that are developed in response.

The trustworthiness of our observations is paramount, but so is the plausibility of the inferences we make on the basis of the evidence. Even if there is no room to doubt our data, some conclusions will be judged better supported by the evidence than others. We can't assume that it will always be immediately apparent what justifies certain claims. Here again the idea of distinctive scientific methods seems relevant. What makes the sciences more *rational*, we might assume, is the practitioner's commitment to certain methods: an individual scientist's gender, race, ethnicity, socio-economic background are irrelevant, we're assuming for now, to how they pursue scientific research. As we noted above, however, allusions to the scientific method, or methods, can't satisfactorily answer our concerns unless we are in possession of a convincing account of what these methods are. It makes little difference whether we frame our project in terms of *how sciences work*, or in terms of *what the scientific methods are* – both will confront the same obstacles and be subject to the same criticisms.

Empiricism is not just an approach towards scientific methodology, but a broader theory about knowledge, its sources and reliability. Big differences exist within the empiricist camp, but these won't concern us much. As a general attitude towards the sciences, the empiricist emphasizes data and observations in the kinds of ways I've described, and assumes there are ways of reasoning from those observations to conclusions that are well confirmed. There are thus lots of important details to work out within the broad, empiricist narrative, and important hidden assumptions that need revealing. The general attitude has been developed in different ways. Some have understood the sciences as centrally concerned with detecting patterns from large data sets. Others have instead focused more on the importance of testing hypotheses against controlled experiments. Popper, as we've seen, described the sciences as being fundamentally concerned with falsification rather than gathering evidence in favour of scientific claims. Most philosophers, however, have maintained that scientific conclusions can become increasingly well confirmed by evidence.

Part of the appeal of the empiricist attitude towards the sciences is its provision of a very familiar and plausible foundation for scientific claims. Our sensory experiences furnish us with oodles of information at any given moment. I make new observations to help confirm or falsify tentative beliefs. I suspect the rice is fully cooked, so I taste a little to convince myself. I doubt the time described by my watch, so I look to other devices to ensure I won't miss my appointment. Gathering new information, modifying or rejecting old beliefs in response, or adopting entirely new ones, is all so familiar that we're seldom even aware of it. Furthermore, it *seems* to work. The suggestion that sciences extrapolate from this homey picture is appealing. There are, however, reasons to worry about straight, naïve or simple empiricist attitudes towards science. Over the course of the next two chapters, we'll address three general issues.

The first draws attention to the fallibility of our powers of perception and observation. Insofar as the reliability of our scientific theories seems to rely on the reliability of our observations, then problems with the latter undermine confidence in the sciences. Interestingly, some problems with our observations can't be overcome by just looking harder or replicating studies, and the consequences of this fact include important lessons for our understanding of all scientific inquiry. A second general concern with empiricism arises when we press the empiricist for more details on how much confidence we should have in hypotheses, and how we should respond to instances where more than one hypothesis fits available data. We will see that *certainty* is a hopeless standard by which to evaluate scientific conclusions, but that understanding what weaker notion is appropriate is not straightforward.

The first two concerns can be thought of as *internal* to the empiricist perspective. Empiricists value data and logic, and so they invite demands for further clarity on both fronts. A third kind of challenge arises when we adopt alternative approaches to understanding the nature of scientific inquiries. As we've already noted, the sciences have their own histories, which are replete with heroes, seminal events and discoveries, as well as embarrassments and disappointments. During the middle decades of the twentieth century, the history of science emerged as an important scholarly pursuit in its own right. These studies suggested to some that the basic empiricist picture must be wrong. Empiricist theories lead to certain expectations about how sciences have evolved over the last few centuries, and to the extent that those expectations don't fit the actual historical record then the empiricist attitudes become less plausible. But it is not only history that threatens empiricism. Some scientific research projects receive generous funding, others none. Some research projects might be considered socially or ideologically problematic, thus are less enthusiastically pursued. One attitude towards such factors supposes that they complicate the way the sciences in fact progress but are little more than curiosities, or perhaps interesting distractions from the main story. More radical use has been made of these sociological considerations, however. According to some, the only way to understand sciences is by recognizing them as a reflection of a broader economic, social and political story.

As we continue to explore the nature of scientific inquiries it will become apparent that they are fallible, a concession that is quite consistent with the sciences still representing our most reliable means of investigating the world. Sciences are processes – not only are our theories modified, but also our methods, our means of evaluating and some of our most deeply held convictions. Sciences can be defined institutionally, with their own history, social structure and reward system, and also in egalitarian terms, where in principle at least everyone's observations and arguments deserve attention and consideration. Gathering evidence, explaining, predicting, testing and revising are central to scientific research, but we should be wary of oversimplifying our description. These various connotations can pull us in different directions, complicating our hopes of utilizing scientific knowledge in the most fruitful way. A better feel for these issues is worth pursuing. In this chapter, I've suggested that no clear solution to the problem of demarcation has yet been offered, questioned whether resolving the problem is as important as evaluating new ideas via whatever methods seem appropriate and, finally, introduced the idea that data, observations and distinctive methods occupy a special place in scientific research. It is time now to start thinking more carefully about the role of data and the forms of reasoning.

Discussion questions

1. Which seems more helpful, to regard all sciences as employing a single method or to suppose that the sciences employ a wide variety of methods, some of which share some prominent features in common?
2. Are we inclined to suppose that the sciences *must* share something in common? If so, what might explain the inclination?
3. Is there a difference between judging whether some study qualifies as scientific and judging whether that study generates reliable conclusions? If so, which seems more important?

4. Critics of certain scientific conclusions sometimes complain that the evidence or method *isn't scientific*. How might we respond to such charges?
5. In some legal cases Creation science has been banned from public science classrooms on the basis that it is not part of the sciences. Could judges instead rule that we shouldn't be teaching *bad* science? How important are the *motives* of those who push for the inclusion of a particular approach in science curricula?
6. I described the absence of a good solution to the problem of demarcation as suggesting that, collectively, we are obliged to consider and evaluate *all* ideas, but was I guilty of exaggeration?
7. How much confidence was it appropriate to place in Mendel's conclusions?

Suggested reading

There are several very good introductions to the philosophy of science which overlap with the first part of this book. Chalmers (2013) is now in its fourth edition and has become a modern classic in philosophy of science. Barker and Kitcher (2013), Godfrey-Smith (2003), Okasha (2002) and Ladyman (2001) are also excellent. Bowler and Morus (2005) is an excellent survey of the history of science. Laudan (1983) is an important paper on the problem of demarcation. For recent work on the problem, see Pigliucci and Boudry (2013).

2

Two challenges for the naïve empiricist

The empiricist attitude places a heavy emphasis on the role of observations within the sciences, sensory experiences and, most generally, data. Sciences are in the business of carefully gathering data, which motivate hypotheses and ideas about the world, which are further explored, tested and investigated by gathering more data. Those hypotheses that are judged implausible, in light of available evidence, are discarded; those hypotheses that are judged acceptable are retained, and their implications further explored. Observations motivate hypotheses and theories, for the empiricist, and provide the crucial means by which we evaluate them. In this chapter we'll explore two central aspects of the very appealing empiricist story. First, if hypotheses are rejected or accepted on the basis of the observations we gather, then the reliability of these decisions clearly depends on the reliability of the observations. Rejecting a hypothesis because it doesn't fit my observations would be bad practice, if those observations are made in bad light for example. Reasons to worry that our sensory experiences are unreliable means of assessing scientific hypotheses is thus simultaneously a challenge to the merits of the hypotheses themselves; if we rely on data to sustain scientific objectivity, then cracks or weaknesses concerning the objectivity of our data will undermine confidence in the sciences more generally. In the second half of the chapter we'll see that favourable evidence can't *guarantee* that our theories are true, but articulating the nature of a more modest attitude towards scientific ideas doesn't permit of any easy formulation. The possibility of observational error and the uncertainty of scientific conclusions each complicate the basic empiricist picture that can seem so alluring. Recognizing these complications should help produce a more realistic and thereby more helpful image of how the sciences work.

Observation as theory laden

That our observations are fallible is utterly familiar. We notice someone across the street, identify her as a friend, wave frantically to attract her attention, only to realize that it wasn't who we thought it was. Whoops. The fallibility of our observations isn't

Figure 2.1 Old woman/young woman illusion

always a product of haste, however. Alan Chalmers describes how naked-eye observations of Venus produced a significant challenge to Copernicus and his heliocentric theory of the solar system.[1] The theory predicts that the distance between Earth and Venus will vary significantly as each orbits the Sun, and consequently that the size of Venus, as viewed from Earth, will also vary. The difficulty for heliocentricism was that centuries of astronomic observation had produced no evidence that the observed size of Venus does change. Only with the invention of the telescope was the prediction confirmed, several decades after Copernicus had died. Copernicus was right about the apparent size of Venus, but for over half a century observational data suggested otherwise. The example is a striking illustration of how that which we might assume is an observable *fact* is not necessarily immune to change.[2] The example also illustrates how human perception is prone to systematic errors. Naked-eye observations of Venus did not provide reliable evidence about its apparent, relative size.

Optical illusions provide another familiar example of systematic perceptual error. Given certain stimuli, we incorrectly perceive two images as different in size, or colour, or brightness, for example. Lines appear as if they're converging, but aren't; part of an image seems to be in motion, but isn't. Auditory illusions are less well known, but no less fun. The Shepard scale is a sound that we hear as continuously ascending and yet always remaining within a fixed range. If that sounds impossible, there are examples online that should convince you otherwise.

In addition to such illusions, psychologists have discovered that our perceptions of ambiguous pictures, like the one above (Figure 2.1), are susceptible to certain kinds of manipulation. The image can be viewed as a young woman or an old woman, but how subjects perceive the image is influenced by, for example, the image or concepts they

[1] Chalmers (2013).

[2] Chalmers (2013, 17) describes the situation as follows: 'we now know that the naked-eye observations of planetary sizes are deceptive, and that the eye is a very unreliable device for gauging the size of small light sources against a dark background. But it took Galileo to point this out and to show how the predicted change in size can be clearly discerned if Venus and Mars are viewed through a telescope.'

are exposed to immediately prior. Changes in context, our motives and emotional state, as well as cultural upbringing, can all affect perception. Our experiences and our observation reports can also be influenced by *expectations*.

Dan Ariely is a psychologist and behavioural economist who conducted an experiment in a Boston bar that illustrates the influence of expectation over experience.[3] Having secured the consent of the barman, Ariely and his collaborators offered patrons a free beer. Customers were given small samples of two beers and offered a free pint of whichever they preferred. One of these samples was unaltered, direct from the tap, as the brewer intended. The other had been doctored – two drops of balsamic vinegar added for each ounce of beer. Hundreds of patrons took part in the study, around half of whom were made aware of which beer contained the vinegar before they sampled either, while the remaining subjects were oblivious even to the fact that any tampering had occurred. The reactions of the two groups were markedly different. Among those who were forewarned, there was a strong preference for the untainted beer. Those ignorant of the meddling, however, generally *preferred* the beer with vinegar added. We expect that adding vinegar will give the beer a less pleasant taste; it is this expectation that seems to explain the attitudes of Ariely's subject groups. The idea that our expectations can shape our experiences, at least in certain respects, has been widely established.

The role of expectation in experience has been used to challenge the objectivity of scientific reasoning. The most important of these considerations – that observations might be theory laden – concerns the possibility that the scientific theories we embrace might themselves influence our sensory experiences.[4] To see how this could lead to problems, suppose different scientists accept different explanations for some phenomenon, where the competing explanations rely on competing scientific theories. N.R. Hanson was one of the first philosophers to explore the possibility that observations are theory laden; he asks us to imagine a geocentric astronomer and a heliocentric astronomer sitting on a hillside at dawn.[5] The geocentric astronomer believes the Earth is fixed and thus the appearance of the Sun rising above the horizon is created by the Sun orbiting about the Earth. The heliocentric astronomer believes the Sun is fixed and hence the appearance of the Sun rising is created by the Earth's spin. Hanson's concern is whether the two astronomers *see* the same thing as they watch the Sun at dawn. He argues that, in an important respect, they do not.

Empiricist philosophers, at least at one time, assumed that the plausibility of an observation report was independent of an individual's theoretical commitments, but

[3] The experiment and results are described in Ariely (2010).

[4] The idea that observations are theory laden, and the problems this ostensibly creates, has been much discussed since the 1960s. Not everyone has understood the issues in the same way, however. Some have focused on the impossibility of describing our observations without using the *language* of scientific theories. Others have been concerned with how expectations influence where scientists look and what they look for. The concept also became important within certain debates in the philosophy of mind. Here I'll focus on what seems to have been the most common understanding of the argument, and certainly the most relevant to our concerns.

[5] Hanson (1958).

work in psychology, like Ariely's, suggests this might not be true. If having distinct theoretical beliefs about the world can produce different experiences, then appealing to observations might be considerably less effective at indicating which of competing theories is most sensible. If proponents of geocentricism and heliocentricism cannot observe the same things, when it comes to making observations that could help resolve their theoretical disagreement, then it appears that observations cannot play the objective role within the sciences that empiricists typically suppose. That observations are theory laden suggests that they may not provide the theory-neutral arbiter between theories that we typically suppose they do.

The possibility that our observations are shaped by the scientific theories we accept is certainly intriguing. The possibility that observations are thereby ineffective, or at least compromised, as means of testing scientific hypotheses is troubling. Human perceptual systems clearly involve more than just the simple reception of incoming sensory experiences. The information that's contained in the retinal image, for example, is insufficient for many non-perceptual cognitive processes; vision is sometimes described as involving an unconscious *inference* based on information from the retinal image and certain unconscious assumptions about the world. For example, the image below is ambiguous. Perhaps it's an image of a light grey rectangle with a darker grey rectangle in front and partially obstructing it, but it could be a dark grey rectangle with a light grey L-shape, positioned off the bottom left corner. Most people report seeing two rectangles, one in front of the other. That many people describe themselves as perceiving two overlapping rectangles can be explained in terms of an unconscious inference, perhaps based on our previous experiences that images like this are typically produced by overlapping objects. But do any of these conclusions about human perception provide good reasons to suppose that different *scientific* beliefs might alter how individuals experience the world? And, if so, have we thereby discovered a distinctive challenge to the reliability of our scientific reasoning?

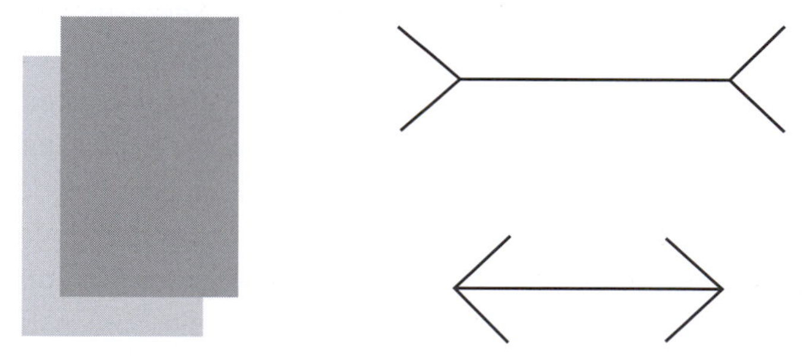

Figure 2.2 Left: Partially obstructed rectangle or light grey L-shape? Right: The Müller-Lyer illusion.

One reason to be suspicious of the idea that observations are theory laden was beautifully articulated by Jerry Fodor.[6] Fodor focuses his discussion on optical illusions. He recognizes such illusions as products of how we process visual stimuli, where these processes involve assumptions about the world. We're unaware of employing such assumptions, but it's these that lead us to perceive the horizontal lines in the Müller-Lyer illusion (figure 2.2), for example, as differing in length. What Fodor then notes, however, is that these illusions persist, even once we've fully convinced ourselves that a given effect is illusory. The illusions are importantly robust to at least some very pertinent changes in our background beliefs. Even once we've measured the two lines, and accepted that they are the same length, we still perceive them as different. Our perceptual experiences are influenced by some background beliefs but not others. The significance of this conclusion is that it now becomes unclear whether working with different scientific theories will have any important influence over our observations.[7] Do scientific theories and conclusions function more like the belief I reach by measuring with a tape measure the two horizontal lines in the Müller-Lyer illusion, (figure 2.2)? That is, a belief that doesn't influence my *perception* that the lines are of different lengths. Or do they function more like the belief that vinegar will spoil the beer, which does seem to affect sensory experience? If the latter, does this create a distinctive challenge to the purported authority of scientific conclusions?

Before addressing these questions it is worth reminding ourselves that our observation reports are always fallible. There is always a risk that we will make mistakes. We can strive to overcome errors by insisting that results be repeated, checked, reviewed and so on, but the risk can only be managed and not entirely removed. We'll consider some implications of fallibility in the second half of this chapter, but for now let's notice that for theory-laden observations to provide a distinctive challenge to our scientific conclusions, they must gesture to more than mere fallibility. One interesting way of achieving such excess requires evidence that holding different scientific beliefs *affects* the way that scientists observe the world or report their observations, and that consequently their ability to properly evaluate scientific theories is compromised.[8] The implication here is not just that scientists might make mistakes, it's that once in possession of certain scientific beliefs their observations (or reports) become less objective, and hence provide a less objective foundation for testing and evaluating scientific conclusions.

Although better conceived as something distinct from the theory-ladenness of observation, it has also been noted, within discussion of observation in the sciences, that scientific *training* changes the way that individuals experience the world. When

[6] Fodor (1984).

[7] Fodor argues that they do not, and hence that observations are, typically, objective and theory neutral.

[8] There is clearly a difference between what people in fact perceive and what they report having perceived, so it is conceivable that embracing different scientific theories might influence either observation reports but not perceptions, or vice versa. In either case, however, the challenges to scientific objectivity are largely the same, so I'll treat these alternatives as one single problem.

students peer through a microscope they might struggle to identify certain features that a more experienced microscopist has no difficulty discerning. If training and instruction change the way individuals experience the world, then we might wonder again whether observations can provide the kind of objectivity for scientific reasoning that we desire. However, granting that scientific training generates different experiences creates a challenge to scientific authority only if such training either produces in scientists an ability to observe what isn't really there, or degrades their capacity to observe what the layperson sees. Neither of these consequences seems likely. Chess grandmasters see sequences of moves that less experienced players don't notice. X-ray technicians look at X-rays and see scars, tumours and infections, while patients see meaningless splotches and lines. In these and many similar cases, we don't doubt that the experts are more accomplished than the layperson at seeing what's actually there. The fact that someone has been trained to use technical instruments and apparatus is a poor reason for doubting her observation reports. Nevertheless, it remains conceivable that the expectations some scientists form, on the basis of scientific commitments, could influence experiences. A purported instance of observations being theory laden in this sense occurred in the seventeenth century.

Galileo Galilei didn't invent the telescope, but he made important improvements to its design and spent a lot of time observing celestial bodies. He described mountains on the Moon, solar spots, the phases of Venus and the moons of Jupiter. The planet Saturn likewise had surprises in store: when Galileo directed his telescope in its direction he described how 'to my very great amazement Saturn was seen to me to be not a single star, but three together, which almost touch each other'.[9] For fifty years astronomers made similar, although not identical, reports. Some described one larger body with two smaller bodies, one on each side. Some described handles on the large body, rather than distinct but proximate bodies. For others, Saturn presented itself as a single body rather than three, an appearance that Galileo himself admitted to having observed on some occasions.

The various observations of Saturn produced for astronomers a genuine puzzle as to what Saturn was. Several decades after Galileo's first telescopic observations of Saturn, Christian Huygens turned his telescope towards the planet. It was Huygens who first described a planet with rings encircling it.[10] Other astronomers soon came to agree – they now saw rings where once they had seen handles or spheres. For those who would offer this as an illustration of observations being theory laden, it was the *novelty* of Saturn's rings, relative to astronomical theories of the day, which prevented

[9] Quoted in Shermer (2012).

[10] The argument that Huygens provided for his own interpretation of Saturn's form was not based solely on his own observations, but on the many observation reports that had been recorded over the preceding decades. For example, Huygens argued that if Saturn had rings then it would appear different from different points on Earth's orbit. Sometimes the angle of the rings, relative to Earth, would make them impossible to observe; at other times they would be very prominent, as viewed from Earth. Huygens thereby offered an explanation for why astronomers had gathered such a range of observations over the preceding decades. For more details see Shermer (2012).

their discovery because astronomers were not primed to see them. Huygens changed the way astronomers perceived celestial objects by introducing a novel, theoretical alternative.

As an example of the influence that theories have on observations, however, the case of Saturn's rings has limited value. As a challenge to the objectivity of scientific conclusions, the example is even less convincing. Although no-one before Huygens described having observed rings, the suggestion that Saturn had handles was quite common. Handles, however, were novel just as rings were, so it is false to suggest that astronomers were severely limited in their capacity to observe something new. Second, many had described Saturn as consisting of three distinct bodies. Huygens was able to demonstrate that this was an optical illusion, the result of inferior telescopes. The three-body observations no doubt played an important role in the confusions surrounding Saturn, but these are more properly attributed to deficiencies with the apparatus rather than astronomers' observations being theory laden.

Perhaps the most important response, however, concerns the fact that these observations were stretching the *limits* of what could be reliably observed with available telescopes. We know to treat such reports with greater suspicion and care, precisely because such observations are stretching these limits. Analogously, I know to place less trust in my ability to identify someone who is standing a quarter mile away over someone who is by my side and in good light. Distant objects, fast-moving objects, very small objects and so on are, under certain circumstances, vulnerable to alternative interpretations. Distinct theoretical commitments might conceivably influence such interpretations, but this is not a good reason to suppose that all, or even many, observations are influenced in this way, just as my inability to identify distant objects with the naked eye doesn't threaten the objectivity of my observations of items that are a few feet away and in good light. The possibility that theories are less well confirmed than we'd assumed because theoretical commitments influence observation reports isn't well supported by examples of observations that are fuzzy, unclear or ambiguous. If we had reason to suppose that much scientific data were ambiguous, then matters would be different, but this isn't the case. Perhaps theoretical commitments cloud the judgements of some scientists when they're gathering observations at the limits of technology, but this provides no general reason to doubt the reliability of scientific data.

A second putative illustration of theories affecting experiences, in ways that are potentially problematic, occurs when different data emerge from different labs. The reason for the differences might be unclear. It is at least conceivable therefore, that they arise because members of different laboratories have different expectations concerning the outcomes of certain experiments, which might in turn be attributable to different theoretical commitments. Even admitting these possibilities, however, there again seems little motivation to regard such instances as a serious and compelling challenge to scientific authority, principally because such disagreements are typically brought to relatively swift and uncontroversial conclusions. For example, in the 1920s Ernst Rutherford and Hans Pettersson were overseeing different laboratories, but conducting similar experiments that involved bombarding various elements with radiation.[11] The reports from the two labs differed. To try to understand the discrepancies one of Rutherford's colleagues, James Chadwick, visited the Pettersson lab. What Chadwick did at the lab, unbeknownst to Pettersson's assistants but as they were making their observations, was intervene in ways that should have prevented those assistants from observing what they subsequently claimed to observe. The assistants weren't trying to deceive anyone, but Chadwick's interventions revealed that the assistants were being influenced by expectations; they reported observing what they expected to see, rather than what their theory predicted they should have seen. Chadwick's experiment – that treated the assistants as its subject – provided compelling evidence that the data from Pettersson's team were unreliable.

The example of Chadwick and Pettersson illustrates again that expectations might influence scientific observations, but simultaneously suggests that the possibility is not itself particularly troubling. Chadwick changed the terms of the experiments in Pettersson's laboratory in response to his suspicion that expectations were influencing observation reports. Pettersson's mistake was with his experimental design. An easily improved design could have, and did, overcome the problem. More generally, once we become aware of shortcomings in our methods, these can be addressed through improved designs, better equipment or more exacting standards. The revised methods won't confer certainty on the data they generate, but certainty isn't an appropriate goal for the sciences anyway.

Importantly, sciences make progress not just by collecting more observations, but also by learning how we observe, where we sometimes go wrong and how experimental design can be improved. There is a long history of experiments being corrected because we learn that some of the assumptions made in an earlier experiment were false, or that certain sources of interference might have been present, or that the methods of detection were too crude. If we ever hoped that much of our scientific knowledge could be justified on the basis that it is derived from infallible scientific experiments, then history puts an end to those hopes. It is entirely reasonable, however, to suppose that we have greatly improved our methods for gathering reliable data and that we are continuing to do so. In particular instances, criticisms of methods may be entirely justifiable and appropriate, but we have unearthed no general reason to regard scientific with suspicion.

The suggestion that scientific theories cannot be well supported by observations, because experiences or observation reports are contaminated by the very theoretical commitments under review, is not well evidenced by history. Mistakes can occur, and some scientists may express more confidence in their observations than is appropriate, but these are just reminders of fallibility. Expectations based on theoretical

[11] The example is discussed in Bogen (2014).

commitments can influence observations, but we've seen no reason to worry that these threats to objectivity can't be straightforwardly overcome through improved experimental design. Beyond these concessions there are three lessons worth drawing further attention towards. First, it is important to remember that our experiences of the world are subject to systemic error. Experiencing the world is not entirely passive, whereby we simply absorb unfiltered, uninterpreted facts. Although often not apparent to us, our subjective observations are sometimes unrepresentative of how things are, and this can produce an important source of error and misinterpretation. Furthermore, many of these inherent biases are not overcome just by looking harder, or closer, or seeking a second opinion. The nature of such biases prevents us from conquering them in straightforward ways. We must learn to accept that sometimes our experiences might be misleading, no matter how convincing they feel.

The second lesson was illustrated by Chadwick's work: recognizing deficiencies in our methods for gathering data is not cause for alarm but an opportunity for progress. It is an opportunity to find better ways of gathering data, for example. A trait of many scientists is a desire bordering on obsession to ensure that the data they gather are as complete, accurate and free from error as can be achieved given the resources available. Of course, not everyone is as careful, which brings us to our third lesson. Since data can be more or less reliable, people may disagree about the value of certain studies, experiments or observations. The quality of the data is hugely important for purposes of evaluating any scientific hypothesis or conclusion which appeals to those findings. Understanding why some studies are dismissed is often extremely pertinent to understanding the state of a particular debate. The inadequacies of certain studies, however, might initially be apparent only to those with appropriate experience and education. A scientific community's rejections of certain studies may indicate critical problems with the methods by which data have been gathered, for example, even if sceptics protest that the rejection is a result of institutional dogmatism.

These three lessons deserve further illustration, and the placebo effect is a well-known phenomenon that should do the job. The effect describes how individuals will often report favourable responses to ailments, pain and other symptoms of illness, despite having been prescribed substances that are pharmacologically inert, or subjected to procedures that couldn't have any benefit. People may feel that their symptoms are improving if they have reason to think that their symptoms *should* be improving. The placebo effect is culturally specific and multi-faceted. Studies suggest that people's experiences can be influenced by whether drugs are advertised as expensive or cheap, their colour, as well as the packaging.[12]

The placebo effect reveals that our personal experiences with a given drug, for instance, are an unreliable mark of its efficacy. Our conviction that a certain medication, or procedure, or therapy helped alleviate our condition might be overwhelming.

[12] For an accessible and entertaining discussion of the placebo effect see Goldacre (2010).

We might find it almost impossible to doubt what we think we know about why our symptoms are so much improved. The placebo effect cautions us, however, that we could be wrong.[13] Nevertheless, the placebo effect does not prevent the medical profession from conducting worthwhile studies concerning the merits of new drugs and procedures – it simply calls for improved experimental design. In its crudest form, trials look for evidence that new drugs perform better than a placebo.

Of course, the evidence generated by some medical trials might be dismissed as worthless, because the researchers failed to adopt an appropriate control group. Other studies might have additional shortcomings. However, the unreliability of one individual's reports, in light of the placebo effect, is not an insurmountable obstacle. Finally, whether a trial accounts properly for the placebo effect can only be evaluated case by case, but failure to establish that a procedure performs better than a placebo is a good reason for finding very little value in a particular study. In all scientific disciplines, there are protocols for proper experimental design, and these can serve as justification to largely ignore the conclusions of experiments that fail to satisfy the appropriate standards. That the layperson may not always understand why some studies have been discredited is quite consistent with the presence of very good reasons for their rejection. Perhaps none of these lessons appear particularly remarkable, but even obvious truths can have a habit of escaping us in those moments when we are influenced by emotion, desire or availability.

The ways in which our impressions, beliefs, observations and descriptions of the world are produced is extremely deserving of continued study and investigation. Scientists in all disciplines continue to seek better ways of gathering data, which sometimes requires correcting existing techniques. It behoves us to exercise caution with respect to the reliability of our observations and experiences. Our data are fallible, but, as we'll emphasize shortly, certainty isn't the correct standard by which to evaluate scientific claims anyway, hence fallibility is expected. Thus, it seems we can fold worries about the reliability of our experiences and observations into the more general concern of how well supported by evidence our theories are. There is always a risk that we might have made mistakes when generating scientific conclusions, but systematic errors in human perception are just a further potential source, and not an insoluble one. Problems with observations don't create a distinctive challenge of their own.

One final, general issue deserves some discussion. The basic empiricist picture seems to imply that by carefully gathering data we can provide an objective foundation for scientific arguments and conclusions. In light of the considerations described in this chapter, it seems we must also demand reasons to accept that the methods by which we gather data are themselves reliable. However, if data can't justify the

[13] The placebo effect is not the only reason that our subjective experiences are unreliable for purposes of evaluating the efficacy of drugs. Many ailments disappear without any medical intervention whatsoever, and even chronic conditions have cycles of more and less acute symptoms. If our conditions would have improved without any drugs, but the drugs are taken anyway, we may easily confuse causation with mere correlation.

sciences, because sometimes even carefully gathered data are untrustworthy, then how are we to justify our preference for *scientific* conclusions over those conclusions generated via other means? In particular instances we might find convincing ways of conquering biases, prejudices and ambiguous data. We can build a better microscope or telescope. We can insist that observers remain ignorant of what they *should* observe, according to some theory, to protect against expectations compromising the reliability of the observation reports. We can measure lines to determine whether their apparently differing lengths are illusory or not. But now we're just gathering data in a different way. In each case we might feel justified in our preference for the new data over the old, but if the problem is one of justifying *the importance of data*, then we don't solve the problem by simply requiring better data.

The mistake here lies in supposing that sciences require, or that we would benefit from, a universal justification, which distinguishes and privileges the sciences from other ways of knowing in terms of their sensitivity to empirical data. If a hypothesis is scientific so long as it is supported by *some* empirical evidence, no matter what the evidence's merits, origins or method of acquisition, then even outlandish, absurd, foolish hypotheses will qualify as science. Scientific authority becomes worthless if the sciences are understood so broadly. The differences between reliable and unreliable data are things that we learn, and that we're continuing to learn. Consequently, there would be little value defending sciences on the basis that they utilize whatever we currently regard as reliable data, since tomorrow's data could be better; what we value, foremost, is the progress achieved, rather than the putative scientific status that both the old and new enjoy. Here we find ourselves reinforcing conclusions suggested in the last chapter concerning the problem of demarcation: rather than look to defend all sciences on the grounds that these satisfy a particular image of what scientific evidence requires, we should insist on more detailed defences that inform us, among other things, of the kinds of evidence that are being utilized in particular cases, and the reasons they can be trusted and preferred over the alternatives.

The problem of induction

We have focused in this chapter, thus far, on the possibility that scientific authority might be challenged by drawing attention to systemic problems with our sensory experiences. We've seen, among other things, that, first, avoiding error is often more complicated than we might expect; second, we should continue improving the reliability of our data-gathering methods; third, we shouldn't trust empirical studies that are insufficiently alert to known problems; and, fourth, scientific conclusions are always fallible. Fallibility, importantly, is consistent with both having corrected prior mistakes and being very good judges, if imperfect ones, across a wide range of circumstances. It remains to be seen how well our image of the sciences survives the observation that scientific conclusions are fallible. An important perspective on questions of fallibility is provided by another famous philosophical problem – the problem of induction.

This morning I washed my hair with shower gel, believing this would leave my hair cleaner and smelling of a fragrance that the manufacturer has named *Phoenix*. I walked downstairs, confident that no floorboards would collapse under my weight, drank orange juice without pausing to consider whether it might make me violently ill and turned the car key in the ignition to start the car. I ate, groomed, and moved around; interacted with people, appliances and buildings. I achieved a great many utterly mundane ends, and I did it all – to a very large extent – by assuming that things would behave today as they've behaved in the past. I assumed that *previous* experiences, with shower gel, stairs, juice, cars, toasters, doors, roads and laptops, were a reliable guide to how they would behave today. I assumed that today's experiences would resemble past, similar experiences, at least in important respects.

The problem of induction is the problem of justifying this very innocent looking assumption. It has been a perennial irritant for philosophy ever since David Hume argued in the mid-eighteenth century that we have no rational basis for inductive inferences. It is a problem that has significant implications for our understanding of how sciences work, and it really is a beast of a problem. Inductive reasoning is essential both to the sciences and everyday living, but there are compelling arguments that suggest that a solution to the problem is impossible. With no solution, we seemingly lack justification for a type of inference that we can't avoid, that feels incredibly natural to us, and for which it's just hard to shake the conviction that *some* kind of justification must be possible. The philosopher C.D. Broad described induction as the triumph of the sciences and the scandal of philosophy. To better appreciate the scandal, we should first say something about arguments, and reasoning, more generally.

Arguments here can be understood as attempts to convince people of particular ideas or attitudes. Evidence, or reasons, or premises are advanced, and, on the basis of these, we're urged to either *infer* or *deduce* a particular conclusion. Arguments can fail to convince us for two *kinds* of reason. First, we might be suspicious about the reliability of the evidence itself. If someone argues that Brazil will win the World Cup this year because it has the most talented individual players, we might object that actually Brazil *doesn't* have the best players. If some of the premises in the argument seem implausible (for example, that Brazil has the best players), then the argument won't convince us (i.e., we won't be convinced that Brazil will win the World Cup *because* it has the best players). Of course we might have quite different reasons for accepting the conclusion, but our present concern is with evaluating *arguments*. An argument succeeds only if it provides good reasons for accepting its conclusion, not just because it contains a conclusion that we find plausible. In this instance our sceptical response is that the reason offered for supposing that Brazil will win the World Cup is dubious. We're inclined not to grant the premises.

The second way in which an argument might fail is if the evidence, independently of *its* plausibility, is irrelevant, or of questionable relevance, to the merits of the conclusion. Brazil is the largest Portuguese-speaking country in the world, the world's largest producer of coffee, and shares borders with all other mainland South American countries except for Ecuador and Chile. None of these provide any reason to suppose that Brazil will win the World Cup. The problem is not that these claims are false, but that they're irrelevant. Even if we grant the premises, we baulk at granting an inference *from* these *to* the conclusion. Recognizing whether premises are relevant for purposes of evaluating a given conclusion is often more subtle, but key to evaluating any argument.

From arguments in general we now turn to a particular species of argument. A deductively valid argument is one for which the truth of the premises guarantees the truth of the conclusion. Deductively valid arguments are sometimes described as truth-preserving, because *if* the premises in a deductively valid argument are true, then the conclusion must be true. Here are two examples of deductively valid arguments:

1) All men are mortal	1) If Brazil has the best players, then it will win the World Cup
2) Socrates is a man	2) Brazil has the best players
3) Therefore, Socrates is mortal	3) Therefore, Brazil will win the World Cup

As is probably apparent, if each of the premises in either of these arguments is true, then the conclusion of that argument must also be true. Or, equivalently, if the conclusion of either argument is false, then at least one of the premises in that argument must be false. Establishing whether the premises are true might not be straightforward, but this is irrelevant for purposes of evaluating whether an argument is deductively valid. Assessing whether a given argument is deductively valid always attends solely to the question of what follows if we *assume* the premises are true.

Now let's consider another argument. Suppose you've observed a large number of white swans and have never observed a swan that wasn't white. You might feel some tendency towards concluding that *all* swans are white, including those you haven't observed, whether they lived in centuries past or will live centuries in the future, whether they live just down the road or on the other side of the world. However, the fact that you've never observed a swan that wasn't white is clearly no *guarantee* that there are no non-white swans. Your observation reports of white swans might all be accurate and true, but they don't assure with necessity the conclusion that all swans are white. Arguments that are advanced on the basis of our past experiences are not deductively valid if their conclusions range over instances that we have not observed.

We'll use the phrase *inductive reasoning*, or *inductive argument*, to denote those cases when we use our past observations as evidence for conclusions that extend beyond the observations themselves.[14] Concluding that the car will start *this* time because it started on all previous occasions that I turned the key in the ignition is an example of inductive reasoning. Similarly for the inference that all swans are white because I've observed only white swans. Scientific reasoning frequently involves inferring to conclusions that range over events, or phenomena, that we have not witnessed directly. Scientific theories are invoked to ground predictions about the future, or predictions about past events that no-one was around to observe. Theories might include generalizations that purport to describe events across all space and time. Newton's law of universal gravitation describes how *any* two massive objects will attract one another via a force that is proportional to their masses and inversely proportional to the square of the distance between them. The sciences seem to rely essentially on inductive reasoning, that is, inferring on the basis of past experiences facts about things that we haven't observed and perhaps will never observe.

Sometimes scientific reasoning might closely resemble the kind of inference illustrated above, from particular observations (of white swans) to generalizations (about all swans), but many scientific arguments seem to have a different kind of structure. Evolutionary biology draws evidence from the fossil record, molecular biology, comparative anatomy, the geographical distribution of species, and more. Collectively, the evidence provides a powerful argument in favour of the central ideas of evolutionary biology, but it doesn't easily fit the model of gathering instances and inferring universal claims. Nevertheless, these arguments still rely on the assumption that instances of which we have had direct experience are a reliable guide (in at least some circumstances) to instances that we have not experienced directly, and this is the assumption that Hume's sceptical argument challenges, a challenge we now turn to.

What Hume recognized is that not only is there nothing inconsistent about supposing that the future could be quite unlike the past, but we have no non-question-begging reason for thinking the future *will* be like the past. (To say that we have no reasons that don't *beg the question* means that the only reasons we do have for assuming the future will be like the past either implicitly or explicitly *assume* that the future will be like the past, and hence they don't solve Hume's problem.) More generally, argued Hume, it is false that 'instances of which we have had no experience must resemble those of which we have had experience'. This might seem entirely unexceptional in cases like that of the swan or turning the key in the ignition. Things don't always behave as they have in the past. But Hume intends his challenge to include cases where we do feel very certain that our experiences provide reliable evidence.[15] Why do I suppose that

[14] These phrases are sometimes used more narrowly, to describe arguments from particular instances to generalizations. The latter are sometimes distinguished from inferences that seem to make essential use of a theory's capacity to *explain* various phenomena. I'm sympathetic to drawing this distinction, but for present purposes it will be more straightforward to regard all arguments as either deductive or not. Those that are not deductive will be understood as inductive.

[15] Russell (1912) includes the following, memorable illustration of the problem of induction: as a result of its past experiences a chicken will come to expect food when the farmer approaches each morning. On one fateful day,

unsupported objects, like rocks, keys and footballs, will fall towards the Earth's surface? Why do I presume that removing a baking tray from a hot oven, without the use of oven mitts, will badly burn my hands? Why do I assume that the rain will make the grass wet or that decapitating a person will result in death? With these and many other examples, we feel supremely confident that our beliefs are reliable, but aren't these beliefs also based on past experience?

At this point Hume anticipates one response we might offer: perhaps such inferences are based on *causal reasoning* and perhaps it is our grasp of causal relations that explains and justifies our confidence in certain projections and predictions. Certainly, we do often reason from observed effects to their likely causes, or from known circumstances to their likely effects. If everyone who ate the chicken became ill, and everyone who avoided it felt fine, we might infer that it was the chicken that caused the illness. If I leave a pizza in the oven for too long, I believe this will cause it to burn. Causal reasoning is extremely familiar, but Hume challenges us to consider what basis we have for trusting inferences that appeal to causal relations.

The mere inspection of objects, no matter how careful, reveals conclusions about neither their causes nor effects. Imagine someone who has had no prior acquaintance with either wood or fire. What Hume recognized is that closely inspecting the wood will provide no reason for the naïve individual to infer that if the wood is placed in the fire then it will become charred. Similarly, careful examination of a charred stick reveals no information about how it became charred. We only acquire knowledge of causal relations, concludes Hume, *through experience*.

Certain patterns in nature become so obvious that it feels as if things *have* to behave that way. The unsupported object *has* to fall towards the floor. The rain *must* make the ground wet. Hume argues that the apparent necessity isn't something we can locate out there in the world, however, but actually stems from us. We can't observe and discover causal relations other than by observing that certain types of events are associated with other types. According to Hume, we achieve beliefs about causal relations by noticing that one type of event temporally precedes a second type, is spatiotemporally continuous with it, and is constantly conjoined via experience with the second type of event.

The significance of Hume's discussion is that, since our knowledge of causal relations depends on past experiences, it is ineffective as a justification for our belief that unobserved events will resemble those similar events that we have observed. My belief that the hot baking tray will burn my hands is dependent on the idea that objects will behave in the future as they've behaved in the past. What reason, asks Hume, do I have for accepting that the future will resemble the past? If we'd hoped to answer that question by appealing to the concept of causation, then Hume has now provided a

however, the inference from past experience will prove unreliable – the chicken will approach the farmer expecting food, only to have its neck unceremoniously wrung.

response. Although it seems as if I *know* that heat causes burning, hence I know the effects of carelessly handling very hot objects, the only way we discover any causal relation is through experience, and so causal reasoning is reliable only if the past resembles the future in relevant respects. We can't appeal to the relation between cause and effect to justify our belief that the future will resemble the past, because our reason for trusting those relationships presupposes that the future will resemble the past. We still lack justification for the belief that the oven tray will burn my hands.

A different response to the problem of induction might have occurred to you already. Our inductive inferences have worked in the past, so isn't this a good reason to suppose that they'll work in the future? As a solution to the problem of induction, however, this response seems to beg the question.[16] Whether past experiences are a reliable guide to the future is precisely what's at issue, so we can't assume their reliability. We can't assume that the success of inductive reasoning in the past reliably indicates that inductive reasoning will be successful in the future.

Another common reaction to the problem is to play down its significance. Of course past experiences don't *guarantee* what tomorrow brings, but the absence of certainty should strike us as neither surprising nor troubling. Unsurprising because, as we've already noted, sciences are dynamic, permanently being revised, corrected and updated, and these are changes that couldn't occur if scientific conclusions were established with certainty. Untroubling because we don't require certainty – we'd happily settle for certainty beyond reasonable doubt. Tempting though the response might be, it again doesn't seem to work. Hume's challenge is not aimed particularly at claims of certainty, but whether our experiences can provide *any* basis for predicting future, or unobserved, events. Consider the following argument:

1) The first swan observed was white
2) The second swan observed was white
3) The third swan observed was white

. . .

n) The nth swan observed was white

Probably, the next swan observed will be white

Introducing the concept of probability seems like a promising move. However, just as we can't be certain that the next swan will be white, even if we've previously observed only white swans, nor can we straightforwardly justify the idea that it is *probable* that the next swan will be white. The concept of probability has received a great deal of attention from philosophers, but – to cut a long, interesting story very short – answering Hume's challenge by utilizing probabilities would require the existence

[16] Some philosophers have argued that the solution is defensible once we distinguish *rule circularity* from *premise circularity*. See Black (1954) in support, and Salmon (1967) in opposition, to this proposal.

of objective facts about the probability of particular statements, where those facts cannot be based on observed frequencies. Many statements that include probabilities are naturally interpreted in terms of frequencies. If it's asserted that one in five adults regularly smoke cigarettes, then we understand this as describing the frequency, or proportion, of regular smokers within the population. However, our judgement about the probability that the next swan will be white can't be based on frequencies, if it's to answer Hume's challenge, because then we'd be assuming that past experiences are a guide to future events. We need another way of assigning probabilities, but it's extremely unclear how this is to be done.[17] We can't solve the problem of induction by simply weakening our inferences to say that conclusions are probably true.

Relinquishing ambitions of certainty is an insufficient answer to Hume's challenge, but we should pause briefly to highlight, spotlight, underscore and emphasize that sciences rarely, if ever, trade in certainties. We've already seen in this chapter that the data we rely on are fallible. The problem of induction shows further that, even if somehow our data were incontestable, the kinds of inferences we wish to draw don't generate infallible claims. For both these reasons it is silly to object that a particular theory hasn't been *proven*. No scientific claim is proven. Proof produces certainty, and for all the many triumphs the sciences enjoy their conclusions are never indubitable.

What can confuse this point is that scientists often seem comfortable in talking about scientific proof, but this reveals only an unfortunate ambiguity in the way we use the word. Technically, a proof is something that establishes its conclusion with certainty, given the axioms and definitions provided; often, however, *proof* seems to connote something more like *established to a high degree of certainty*, or *well supported by available evidence*. On the first interpretation, no scientific theory ever achieves proof, and thus complaints that any particular conclusion has not been proven are correct but irrelevant. On the second interpretation, it is often no longer clear that sceptics are correct when they opine that particular results have not been proven.[18] On either reading what's important to remember is that the sciences don't deal with certainties and hence the mere absence of certainty in a particular case is unproblematic. Worthwhile objections to scientific ideas require more than airing a grievance that those ideas have not been proven.

Returning to the problem of induction, we can't defend inductive arguments merely by tempering our ambitions, or appealing to causal relations, but

[17] For more on probability theory and the problem of induction, see Okasha (2002) and, for a more detailed, technical presentation, Vickers (2014).

[18] Clarence Cook Little was director of the Scientific Advisory Board to the Tobacco Industry Research Committee in the 1950s and 1960s. As a prominent voice for the tobacco industry, Little continued to insist that there is 'no proof' that smoking could be blamed for lung cancer. Once alerted to the crucial ambiguity in this statement, however, we can see its irrelevance. Genuine proof is not attainable in science, so its absence in this case is neither surprising nor worrying. If we understand Little as implying instead that there is insufficient evidence supporting the connection, then he owes us a far more careful, detailed argument explaining why his evaluation differed so markedly from the overwhelming majority of physicians and medical scientists. Appealing to the absence of proof is at best so hopelessly vague that it is a worthless complaint.

perhaps there are other ways of diffusing the problem. Absent from our current discussion of the problem, thus far, is the fact that the sciences often involve *testing* hypotheses. Isn't it reasonable to trust our scientific conclusions so long as they've passed an appropriate battery of tests? Well, we should certainly admit the enormous value and centrality of hypothesis testing within scientific reasoning, but it's equally important to recognize that allusions to testing give rise to further, difficult and sensible questions. Suppose our intent is one of testing a particular hypothesis. We are careful with our design, the assumptions we make, and the manner by which we gather evidence. Our target hypothesis passes the tests, but what does this tell us about the hypothesis? How much confidence in its accuracy is appropriate? The most significant complication is that there are always alternative hypotheses that are consistent with the results of those same tests. The evidentiary support that accrues for a given, tested hypothesis depends on the number of alternative hypotheses that would have also passed those same tests. Evaluating the number of such alternatives, however, and the extent to which they might depart from the target hypothesis, may be extremely challenging and must somehow account for the possibility that there are alternative hypotheses which we haven't even conceived of.

Whether we observe, then derive conclusions, or conduct carefully controlled experiments, our foremost concern is one of gathering useful data. It has long been recognized, however, that scientists are motivated by features of theories and hypotheses beyond their mere *consistency* with available evidence. Consider the following toy example:

HYP_1: All swans are white.
HYP_2: All swans born before January 1, 2020, are white and all swans born on or after January 1, 2020, are chartreuse.

Both hypotheses are false, but let's ignore the inconvenience of black swans. Notice, then, that no observations have been made that falsify the second hypothesis. If we've seen only white swans, then until we turn the calendar to 2020 we can have seen nothing that's inconsistent with HYP_2. Patently the first hypothesis strikes us as far more palatable, but what can we say in defence of the judgement? Both hypotheses have untested consequences and both predict that we will have observed to date only white swans. One thing that strikes us as peculiar about HYP_2 is the lack of motivation for that second part, that bizarre appendage or unneeded complexity. Perhaps we prefer HYP_1 because it's simpler.

Many scientists have suggested that *simplicity* plays a role within scientific reasoning. The details, however, are hard to pin down. What does it mean to say that one theory is simpler than another? Why should it matter, for purposes of choosing between competing theories, that one is simpler than another? It's not clear that the *world* is simple, so why should we expect our efforts to describe and understand the world to be simple? Simplicity isn't the only virtue for which scientific theories are lauded. We prefer theories that are broad in scope, are precise and cohere with other well-confirmed scientific ideas. Yet for all their initial plausibility, appealing to such virtues is far from decisive: it is hard to say, in general terms, exactly how we should understand them, why they matter and how we should balance them against one another and against their fit with available data. If we place a very heavy emphasis on simpler theories then we will ignore many outliers (those data points that don't fit a given curve or hypothesis), but ignoring too many outliers will draw sensible complaints that we are ignoring the data. Getting the balance right, between simplicity and honouring the data, permits of no easy, general answer. In particular instances, we might be able to define our terms more carefully and demonstrate how they're relevant to particular goals that we care about. As with our discussion of testing, however, it appears we are pushed towards a more local perspective on issues of evidence and support.

Similar problems plague attempts to contrast theories in terms of their *explanatory* achievements. Clearly many scientific pursuits extend beyond merely *describing* the world we inhabit, its contents, history and regularities, and in addition advance *explanations* for why things are as they are. We don't stop with the conclusion that the oceans are salty; we try to explain why. We try to explain why polar bears are white, why stars twinkle, why acids and bases react and so on. Describing what something is like, or what has happened, is distinct from explaining why it's that way, or why it happened. Straight inductive arguments from observations of particulars to general statements, furthermore, make for disappointing explanations. My impeccable observations of some population might furnish excellent reasons for supposing that all members of the population have some particular property. I have not thereby explained why they have that property or, arguably, why any particular individual does. An appealing means of evaluating scientific theories, therefore, suggests that we should not settle for theories that agree with observations – we should evaluate our theories also in terms of how well they explain our observations. Unfortunately it has proved no easier to articulate what constitutes a good explanation than it is to justify in general terms a preference for simpler hypotheses or more precise ones. Faced with competing theories, we might feel a strong tendency towards identifying one theory as providing better explanations, but it is difficult to describe in general terms what that means, or why it matters.

We lack a convincing, general defence for the idea that hypotheses that have been properly tested should be accepted or that simpler or more accurate or more explanatory theories are always preferable. Consequently, we still lack a helpful response to the problem of induction. We are still without justification for inferences that feel sensible and sometimes even obligatory, and for a style of reasoning that is central to scientific methods. The absence of justification matters. Within the context of a scientific dispute, the concepts of simplicity, testing and explanatory power will provide no assistance if those on either side of the debate disagree about which assumptions are simpler, which tests are most significant or which explanations are better. Conceding that we can't expect certainty within the sciences doesn't answer Hume's challenge either. Nevertheless, combining a more modest attitude towards scientific claims with a more local strategy for justification does, I propose, finally, gesture us in a more helpful direction.

The crux of the problem of induction is that we lack justification for believing that the *unobserved* instances of some kind, type or phenomenon will resemble instances that we have observed. A global defence of inductive reasoning would demonstrate why it is reasonable to suppose that the unobserved instances will resemble the observed instances, a demand I think we simply can't fulfil. However, this sceptical conclusion is consistent with supposing that some observed patterns and regularities are *more* reliable indicators than others of what has not been observed and, furthermore, that we can learn more about which regularities are more stable. We know that some regularities are confronted by more exceptions than others, and often we can explain why this is the case. We know that inferences from samples to populations are less reliable if there are reasons for suspecting that the sample was biased, or is hopelessly small. Any inductive inference from samples to populations presupposes that the observed cases resemble the unobserved in at least some respects, and thus suffers from the fact that there is no *general* defence of this presupposition available. We can nevertheless work to improve the reliability of our inferences by ensuring that our sample is big enough and isn't unrepresentative in ways we can identify.

Inferences from samples are one kind of inductive inference, but the idea of defending particular inductive arguments (or particular kinds of inductive arguments) can be extended further. In 1958, Matthew Meselson and Franklin Stahl conducted an experiment that has become known as the most beautiful experiment in biology.[19] Their interest was the mechanism by which DNA replicates. At the time three hypotheses had been proposed. Meselson and Stahl devised an experiment which quickly convinced other molecular biologists that DNA replicates in the way described by the semiconservative hypothesis: the two strands of DNA molecules separate during replication, with each strand then acting as a template for the synthesis of a new strand. The experiment involved growing bacteria in the presence of a heavy isotope of nitrogen. DNA that's grown in these conditions is heavier than that which is grown in normal conditions. Meselson and Stahl utilized this fact to test competing hypotheses for DNA replication, a test that involved developing a technique for weighing DNA very precisely. There are many good questions we can ask about the confidence biologists placed in Meselson and Stahl's argument. Why assume that all DNA replicates the same way, or that there weren't ways of explaining DNA replication beyond the three hypotheses being entertained at the time? Why accept that Meselson and Stahl's technique for weighing DNA was reliable? What else was

[19] See Weber (2009) for a nice discussion of the experiment.

assumed within the experiment, and how plausible were those assumptions? For the purposes of this book, pursuing these questions in any detail would unfortunately drag us too far from our intended objectives. The more important point, however, doesn't concern the details of this particular experiment, but the idea that the merits of any study can only be properly evaluated by engaging these kinds of detail.

Rather than defend particular inductive arguments by first seeking a global justification for all inductive arguments, scientific claims are likely better defended by attending to the *particular* evidence that's available and the plausibility of whatever further assumptions are being employed. Such justifications will require further evidence, reasons and arguments. Ultimately, we may find that any given argument rests at some point on assumptions that can't be justified, a problem which we'll postpone worrying about for now. Importantly though, an inability to solve the problem of induction is entirely consistent with having arguments which are reasonable given certain assumptions about the world, where those assumptions are judged reasonable given further assumptions and so on.

Hume asks why we should suppose that past experiences provide a reliable guide for predicting future events, but perhaps this is the wrong question to ask. First, we don't always regard observed patterns as reliable guides to future events. That a flipped coin has landed heads four times in a row is not a good reason for supposing it will continue to land heads. The better questions to ask concern why some observed experiences are *better* guides to unobserved instances, what assumptions are being made when we draw these inferences and how plausible those further assumptions are. We shouldn't suppose that answering such questions is easy, but we can suppose that significant progress can be made with these issues when we focus attention appropriately. Sometimes we may discover that our inference patterns require refinement. Sometimes we'll discover that scientific conclusions rely on assumptions that lack significant justification. Conceding that there is no general solution to Hume's problem, however, doesn't entail that we can't advance plausible evidence in support of substantive scientific conclusions, although it does require that defending particular scientific theories will always involve careful engagement with the details.

This attitude towards the problem of induction can again be seen as an extension of the attitude towards demarcation described in the last chapter. There I argued that we are less likely to make progress with respect to the questions that most concern us, simply by reflecting on competing *definitions* of science. Here I want to recognize similar limitations if we fixate on certain features of scientific reasoning. Whether someone claims evidence for a given conclusion, or that insufficient evidence exists, whether people claim that certain hypotheses have been stringently tested, or whether they advance hypotheses on the grounds that they best explain certain phenomena, are never decisive reasons for accepting, or rejecting, scientific conclusions. Data can be more or less reliable, tests can be more or less rigorous, and explanations can rest on more or less plausible assumptions. These are details that matter. The problem of induction doesn't permit of a general solution, but this needn't hinder efforts to evaluate narrower scientific claims.

Our discussion of the problem of induction has by now produced some important conclusions. First, we should not expect the sciences to establish their theories, laws or conclusions with certainty. Second, softening our ambitions and looking for highly probable, or mostly correct, conclusions doesn't straightforwardly answer the problem. (A further problem with this suggestion, beyond those already mentioned, is that of articulating what it even *means* to say that a particular theory is very probably true or approximately true, if these concepts convey something other than the idea that we have lots of evidence that lends credibility to the theory.) Third, and more positively, it is reasonable to hope that we can defend some inductive inferences as much better than others, although these defences will appeal to further assumptions that themselves require justification. What we should say about this possible regress is a subject we will return to in Chapter 4.

Conclusions

The empiricist picture we started with has a powerful hold over us, in part at least because it reduces what can seem difficult and unfamiliar to ideas that readily understandable and commonplace. It's a comforting narrative to suppose that are scientists gather evidence, propose hypotheses, test them and conclude either that target hypotheses are falsified or supported by available evidence. In this chapter we've drawn attention to important complicating factors, important because ignoring them leads to poor evaluations of scientific debates.

The first bedevilling feature concerned potential problems with our observations. Ultimately I argued that philosophical worries surrounding the reliability of our observations should be understood as important illustrations of scientific fallibility, that we should remember that our own experiences don't always accurately represent what's really going on and that the reliability of available data is always paramount. Second, because scientific arguments rely on background assumptions, the plausibility of those assumptions is also always a potential source for diverging evaluations.

Disagreements concerning both the reliability of the data and background assumptions are only settled by advancing further arguments, which themselves might be challenged. It appears highly unlikely that we will settle such disputes by justifying all sciences with one cavalier and comprehensive gesture, but the difficulties associated with trying to identify *general* rules for evaluating all theories shouldn't be confused with a failure to make plausible arguments in particular cases, and even across ranges of broadly similar cases. Perhaps there are no useful and general rules that facilitate reasonable judgements concerning the merits of competing hypotheses across all actual and potential disputes. Perhaps there are general rules, but they're only useful when augmented with particular information that's specific to the domain we're operating within. Just as we have more to learn about human perception, so scientists, mathematicians and philosophers of science continue to study the ways in which scientific arguments work, what justifies them and what assumptions are involved. I suspect most philosophers would agree that there are no magic bullets which discriminate all the good scientific arguments from the bad ones. That still leaves room for good arguments and bad ones, and reasons to suppose that we are getting better at discerning the differences.

Discussion questions

1. To what extent is it a problem that the evidence we appeal to can be of greater or lesser quality? What does this tell us about reasonable expectations for any evidence-based investigation?
2. Is perfectly reliable data ever possible? Why, or why not?
3. Are there general criteria by which we can identify superior data?
4. Are there ways of identifying those occasions when our personal experiences of the world are misleading?
5. Scientists employ a wide variety of rules for inferring, drawing on advanced mathematics, including a lot of statistics, as well as computer modelling. Is it reasonable to expect laypersons to understand all these methods in detail? Does your answer have implications for the level of confidence we should have in scientific experts?
6. It is sometimes suggested that scientific training affects what scientists look for. If true, would this compromise the reliability of scientific arguments and evidence?
7. Is learning more about how we learn a part of the sciences? If so, and since *we* includes scientists, what consequences does this observation have for our understanding of concepts like scientific progress, rationality and objectivity?
8. Do we have a tendency to regard the absence of certainty as reason to treat all ideas as equally plausible? Is this tendency sensible?
9. I can't be certain that my ticket won't win the lottery. Does this mean I am as likely to win as not? How does this example help us think about the absence of certainty more generally?

Suggested reading

Kuhn (1962) and Hanson (1958) were the first philosophers to discuss the problem of observations being theory laden. For a helpful review of the issues, see Bogen (2014). Hume (1738) is still a good source for appreciating the problem of induction. Russell (1912) and Salmon (1967) include good discussions of the problem. For a thorough overview of the current state of the debate, see Vickers (2014). The more positive view on induction I sketched drew much inspiration from Sober (1988), Norton (2003) and Okasha (2005).

3

A revolution in how we think about sciences

The empiricist attitude towards the sciences can appear so very familiar and so obviously correct that it is hard to imagine what an alternative theory of science would even look like. What could be more mundane than the observation that scientists gather data and then reason, or argue, on the basis of those data for certain theories and hypotheses. Surely this is exactly what scientists do, and by doing this they achieve, we suppose, an ever more accurate and complete understanding of the world we inhabit. Admittedly, as we've now seen, even the most basic empiricist ideas face challenges. A naïve empiricist might suppose that our observations are always reliable. They aren't, but the problem of distinguishing reliable from unreliable data admits of no general solution, in part because learning to make the distinction is itself an ongoing process. The problem of induction provides further reminders of both the fallibility of scientific theories and the existence of hidden complexities when it comes to describing and evaluating methods, hypotheses and theories. Nevertheless, these problems don't obviously require the abandonment of empiricism – a little refining might seem more appropriate.

In the first half of the twentieth century, the most influential philosophies of the sciences were empiricist philosophies. Although there were important differences in the details, these accounts shared a commitment to articulate and justify the scientific method, and somehow vanquish the problem of induction. Such articulation, it was assumed, would elucidate what scientists do and explain why sciences make progress. These empiricist philosophers relied on ideas from the philosophy of language, as well as developments in logic, and generally assumed that scientific evidence can be derived unambiguously from observation. They assumed that understanding the scientific method required no serious engagement with the history of science. These were ambitious, extensive and influential theories about how sciences work, but in 1962 Thomas Kuhn's *The Structure of Scientific Revolutions* appeared, and things weren't the same after that.

Kuhn's book revolutionized the ways in which scholars came to think about the sciences. It contained profound challenges to conventional wisdom on the nature of scientific rationality, scientific methods and scientific progress. Kuhn argued that we cannot responsibly ignore the history and sociology of science if our hope is to understand the nature of the sciences. Prevailing empiricist descriptions, Kuhn opined, fail to describe both what scientists now do and what past generations did. Kuhn claimed that patterns in the history of the sciences, which he thought were apparent to anyone who seriously engaged the subject, suggested a very different picture of how the sciences work. Kuhn's work sparked a colossal reaction. Much of it was critical. Kuhn himself was surprised and at times dismayed by some of the ways in which his ideas were interpreted and utilized. What's undeniable, however, is that Kuhn had an enormous and lasting effect on not just the philosophy, history and sociology of the sciences, but on scholars working across an even wider range of disciplines. In this chapter, we'll introduce Kuhn's views, because beyond all the noise and fury that once surrounded them, there are important lessons to learn from Kuhn.

Kuhn's theory of the sciences

Originally trained as a physicist, Kuhn subsequently undertook studies in the history of the sciences. These had a profound influence on the way he came to think about the nature of scientific inquiry. In the opening sentence of his infamous book, Kuhn describes how: 'History, if viewed as a repository for more than anecdote or chronology, could produce a decisive transformation in the image of science by which we are now possessed.'[1] The image which Kuhn supposes could be so decisively transformed is the empiricist one. Kuhn understands the empiricist perspective as generating certain expectations concerning how the sciences should have advanced. These expectations, argues Kuhn, conflict with the actual historical record. For example, an empiricist account might lead us to suppose that the sciences exhibit a *cumulative* pattern, whereby new ideas are accepted on the basis of clear and compelling evidence, and in conjunction with the application of an unchanging and normatively justifiable scientific method. Scientific progress, on this picture, is thus a matter of adding to the store of already existing scientific facts.

Kuhn supposes that many philosophers, scientists and laypersons regard scientific progress as having something like this cumulative bent. However, observes Kuhn, the history of the sciences can't be sensibly interpreted as a story of simply accumulating more facts. In the eighteenth century, scientists adopted Newton's idea that light is composed of material particles. The prevailing view of the nineteenth century was that light is a wave. Our modern understanding describes light as having wave-particle duality, meaning that light exhibits some wave-like characteristics and some particle-like characteristics. As these new attitudes were adopted, they affected how scientists

[1] Kuhn (1962, 1).

interpreted and understood a variety of familiar optical phenomena. Old explanations had to be revisited and revised. The cumulative model must be at least somewhat misleading – simply growing the number of established facts is inconsistent with having to re-examine, re-evaluate and re-interpret the old *facts*.

Of course admitting that sciences aren't straightforwardly cumulative is quite consistent with supposing that sciences still make progress. We just need a different account of how progress is achieved. Perhaps the easiest way to retain the idea, confronted with a history of revisions, starts by admitting that scientists sometimes make mistakes. Over-confidence, prejudice and dogmatism are familiar human traits. These must surely, at least occasionally, disrupt scientific investigations and help generate conclusions that are not fully convincing in light of available evidence. So perhaps the sciences make progress because their practitioners are, for the most part, able to overcome biases and prejudice, although conspicuous and widespread lapses help us explain those occasions when sciences went wrong. Viewed through this lens, the history of science becomes a story of heroes and villains; the heroes recognize certain prevailing beliefs as resting on nothing more than superstition and presumption, rather than evidence and argument. Their personal success in recognizing these facts is attributed to their enhanced objectivity and rationality. Those who oppose the new scientific ideas are regarded, at least with hindsight, as stubborn and irrational, obstacles to scientific progress who failed to follow the rules of proper scientific reasoning.

Kuhn urges, however, that history doesn't square with this revised understanding of progress either. Scientists are capable of prejudice, superstition and irrational commitments, but Kuhn finds no means of separating these from *scientific* attitudes such that we can attribute past generations' triumphs to scientific beliefs and their mistakes to instances of bias. Of course, from our present vantage we will regard some aspects of past theories as correct and others as mistaken. But this is insufficient for us to conclude that the former were the result of correctly applying scientific methods and hence to dismiss the latter as products of irrational tendencies. For Kuhn there is simply no historical evidence of a universal and unchanging scientific method which, once properly augmented with relevant data, yields clear and incontrovertible conclusions. We cannot judge the winners and losers of past scientific debates by appealing to such a method. Rather, scientific methods, attitudes and commitments, have each themselves been the subject of substantive debate and disagreement, according to Kuhn, resulting in new methods and attitudes being introduced and old ones being revised or replaced.

In earlier chapters, we have seen reasons to doubt that there is a universal, scientific method, but in the 1960s this suggestion had profound implications for the philosophy of science. Insofar as the sciences were perceived as models of rationality and objectivity, it was presumed that these qualities were consequences of the application of a distinctive method. If theories are not endorsed because the scientific method recommends them, then why are they adopted? Many interpreted Kuhn as arguing that scientific change was not rational and that its outcomes were the product of contingent historical, psychological and sociological factors. If this was correct, then the idea that the sciences make progress is open to doubt. Whether these interpretations are fair to Kuhn isn't always clear. On any reading, however, Kuhn's work presented an important challenge to many traditional ideas about how the sciences advance.

Kuhn's project was always more than a critique of existing views; he also described a positive account of how sciences function, an account that Kuhn derived from patterns in the history of the sciences which he claimed are common and clear. The central and guiding idea within Kuhn's positive thesis was that scientific work is always conducted within *frameworks* of beliefs, principles, methods and guidelines, which Kuhn called *paradigms*. The practicing scientists of any mature scientific discipline work within a paradigm, or framework.[2] The core assumptions of that paradigm are not challenged by practicing scientists, at least not outside exceptional circumstances, but they are initially pursued and subsequently adopted despite failing to answer prominent problems within the field, as well as containing important ambiguities in the details.

A paradigm includes ideas about what kinds of things there are in the world, their properties and how they interact with each other and our senses. To this extent, a paradigm might not seem much different from our conventional idea of a scientific theory. However, Kuhnian paradigms also provide guidance on how particular sciences should be pursued, what kinds of questions it is appropriate to ask, how one should go about trying to answer those questions and how putative answers should be evaluated. These commitments effectively define science for a given group of scientists and may persist unchallenged for significant periods. Work that is conducted within a paradigm is what Kuhn calls *normal science*.

The paradigms of normal science can be overturned, but for Kuhn such events are quite distinct from the normal course of scientific investigation. Paradigms are only replaced in *revolutionary science*. Understanding the nature of scientific change, for Kuhn, thus requires distinguishing between two very different types of change. We will return to the dynamics of normal and revolutionary science shortly. However, before scientific disciplines even reach the maturity of normal science there often exists, on Kuhn's account, a period of pre-paradigm science. Our more detailed report on Kuhn's views begins there.

Pre-paradigm science – the history of the sciences, for Kuhn, is most naturally understood as a succession of paradigms. Paradigms, and the theories they include,

[2] The concept of a paradigm is the most important in Kuhn's essay. Unfortunately, as he later conceded, Kuhn was not consistent in how he employed the term. The biggest source of confusion was that *paradigm* was used by Kuhn both to refer to the kind of achievement that launched a period of 'normal science' and to a more inclusive set of ideas that describe the shared commitments of a community that practices 'normal science'. The latter sense was a far more atypical use of the word in the 1960s, but it is the sense we will adopt throughout the remainder of the chapter.

are presented, developed and ultimately replaced. Taking a journey back through time thus allows us to retrace the sequences of paradigms and theories that preceded our own. As we travel further and further back, Kuhn suggests, our tours will typically, eventually, reach a period in which there was no generally accepted scientific theory at all, but where something resembling scientific investigation is occurring. Modern textbooks describe light as being comprised of photons, whereas nineteenth-century physicists regarded light as a wave-like disturbance in the luminiferous aether. Newton's suggestion that light is a stream of particles was widely adopted through much of the eighteenth century, but between antiquity and the rise of Newtonian optics, says Kuhn, there were several competing schools of thought concerning the nature of light, but no single view that dominated the others.

Kuhn claims historical evidence for the idea that before the emergence of a scientific achievement, which unifies a scientific community and helps define a period of normal science, there is often a period of pre-paradigm science. Those who operate within such periods are presented with a confusing array of phenomena, similar in certain respects, dissimilar in others. Observations and data alone don't gesture clearly towards one general theory of the phenomena being examined. Consequently, choices must be made about how studies of heat, or electricity, or light, should even commence. Some investigators regard some phenomena as more important than others for purposes of developing a general theory for explaining and understanding. Different methods of investigating might be favoured by some and rejected by others. Distinct schools of thought may also disagree about the appropriate standards by which ideas should be evaluated. Given such diversity among core ideas, Kuhn suggests, there is simply too little common ground from which to launch helpful dialogues between the competing approaches.

Although practitioners within pre-paradigm science might superficially all appear to share broadly similar intellectual interests, their core commitments diverge sufficiently that often they're incapable of appreciating what others are trying to accomplish. The absence of consensus means that individual investigators are each forced to build up theories from ground level. The concepts that are introduced, the methods utilized and choices made by individual investigators about which phenomena to focus on, for example, must all be independently justified. Investigators cannot use a platform of shared knowledge and assumptions to launch new investigations, because there is no such platform. Those from competing schools who read new works will disagree on at least some core points and are therefore likely to regard the work as fundamentally misguided. The absence of common ground between these differing approaches thus creates a serious obstacle to meaningful progress. Kuhn claims the histories of scientific investigations of light, electricity, motion, heat, chemistry and geology all instantiate this pattern. A small of number of competing, incompatible approaches were adopted. Each enjoyed some success, and some schools made contributions that became integrated into the paradigms of normal science, but the overriding impression is one of conflict and disagreement, which in turn obstructs progress.

Normal Science – The shared commitments that are importantly missing from pre-paradigm science emerge, for Kuhn, with the rapid approval among scientists of a particular scientific achievement. Normal science is inspired by, and modelled on, a scientific accomplishment that comes to be judged as an exemplar of how science within that domain should be conducted. The new approach is novel and quickly attracts adherents. It won't explain everything that investigators were previously motivated to explain, nor will it convince all working scientists. The theory will appear particularly successful at accounting for certain types of phenomena and will also appear to hold particular promise with respect to investigating a range of further, related phenomena, and for these reasons appears worth developing and refining.

The arrival of certain inspiring achievements rouses groups of scientists to try to emulate and expand upon the ideas contained within. Scientists who share such hopes will thereby also share important assumptions and attitudes about what they're trying to do and how they should set about doing it. A scientific community that embraces the core ideas that are contained within seminal works will agree on the kinds of things there are in the world, how these things behave and interact, as well as how they can be measured, manipulated and analysed. Notable scientific achievements don't simply defend a new conclusion about the world: they defend, or at least vividly illustrate, a particular way of *doing* science. The architects of such scientific theories thereby set the tone for what kinds of questions it is appropriate to ask, how one should go about answering those questions and how answers should be assessed. Working within a period of normal science involves embracing a wide range of attitudes which are shared by others within the community. Some of the skills and knowledge possessed by practitioners of normal science might not be explicit. The initial scientific achievement that galvanizes a scientific community will include solutions to problems that its author has answered. These solutions will serve as models for future work, so for Kuhn normal science is better understood as utilizing and developing certain forms of scientific practice, rather than adopting and following explicitly stated rules.

Kuhn relates the professionalization of new scientific disciplines and the education of students each to the concept of normal science. The birth of a new paradigm is often quickly followed by the establishment of specialist scientific journals, new professional societies, meetings, university departments and so on. These are a direct response to, and coalesce around, the new approach and the community's shared commitments. Kuhn emphasizes, however, that the new theories are always incomplete and never free from problems; those who are unwilling to accept the new ideas therefore cannot always be straightforwardly regarded as irrational or stubborn. Nevertheless, the growing influence of the paradigm and the growing professionalization of the discipline will push dissenters increasingly towards the margins of scientific discourse. By rejecting core aspects of the new paradigm, they are not doing science as it is now understood.

Meanwhile, the next generation of scientists is being educated within the boundaries of this period of normal science; hence students are taught that doing science involves adopting the practices and assumptions that the community has at least implicitly espoused. Kuhn regards such education as having a very powerful influence over scientists' attitudes, rendering it extremely difficult for individuals to even comprehend alternative ways of pursuing scientific research within their area of specialization. Becoming part of the scientific community, for Kuhn, is a process of learning how to see the world through the lens of that paradigm. Through the professionalization of the community and the education of new scientists, the core ideas of a scientific paradigm, which are initially adopted in part for the promise they seem to hold, come to be regarded as unchallengeable. Kuhn believed his account of normal science is a better description of how science functions. By paying close attention to the history of the sciences, suggests Kuhn, it becomes apparent that some scientific assumptions are, for extended periods, simply not open for discussion and debate.

Normal science requires a community shared framework which is inspired by a particular and notable achievement, but what does a normal scientist do? Kuhn describes several types of work for normal science. For example, facts that are particularly salient from the perspective of a given paradigm are investigated and measured with increasing precision. Some scientists will look for ways to export the methods and techniques that were successfully utilized for purposes of explaining some phenomena into related fields of inquiry that are not yet covered by the paradigm. There is a great deal of work to be done, Kuhn stresses, once a paradigm has been accepted by a scientific community, but importantly scientists working within a period of normal science agree on the kinds of problems they should be addressing. The paradigm focuses the attention of working scientists in a way that involves pursuing questions in far more detail than would otherwise occur.

In describing what happens within normal science, Kuhn also quite deliberately talks about such work as 'puzzle solving'. The unfortunate connotation is that the work of normal science is trivial, but Kuhn doesn't intend that implication. His motivation for describing the work as puzzle-solving is the analogy with familiar types of puzzle. Crosswords, jigsaws, sudokus and the like are presented to us unsolved, but with clear, unambiguous directions for what counts as a proper solution. You haven't solved a crossword, for example, if you've forced more than one letter into a single box. A sudoku isn't solved if the same number appears twice in the same row or column. There are clear ways of identifying failed and successful puzzle solutions. Likewise, suggests Kuhn, when it comes to solving puzzles within a scientific paradigm, there are rules for assessing proposed solutions – rules that are part of the paradigm. Kuhn draws further parallels between the activity of normal science and puzzle-solving. For example, we work on crosswords with the expectation that a solution is possible and that the solution won't involve bending or changing the rules. Kuhn emphasises that normal science does not introduce significant theoretical novelty. Nevertheless, puzzle-solving can be enormously challenging. For Kuhn, working within normal science often 'requires the solution of all sorts of complex instrumental, conceptual, and mathematical puzzles'.[3]

To illustrate some of these ideas, let's return to the cross-breeding experiments of Gregor Mendel that we described in Chapter 1. As we noted, Mendel provided a new approach to questions of inheritance. He focused his attentions on discontinuous traits, crossing pairs of plants that differed with respect to these traits. These led Mendel to propose principles of hereditary. His work was largely ignored for several decades, but in the early twentieth century it very quickly attracted significant attention. Biologists began cross-breeding other species – bean plants, mice, fruit flies and more – hoping to learn whether other traits would fit the patterns Mendel had observed. Quickly there were connections drawn between Mendel's principles and recent discoveries made by cytologists who were developing staining techniques and observing cells under microscopes. These connections suggested an important relationship between cells' chromosomes and Mendel's factors.

At least in outline, Mendel's work exemplifies some key characteristics of a Kuhnian exemplar, and what became known as Mendelian, or classical, genetics seems a fair example of a Kuhnian paradigm. Mendel had proposed that observable traits (like flower colour, seed shape and so on) are determined by *units* or *factors* (which later became known as genes). An individual organism inherits, at random, one such factor from each parent (where each parent also has two factors). If the factors from each parent are the same, the individual is said to be *homozygous* for that trait. If the factors are different, the individual is *heterozygous*. In the latter case the observed trait corresponds with the *dominant* factor. For example, pea plants that received a factor for yellow seeds from one parent and a factor for green seeds from the other always developed a yellow seed; although the plant possessed factors for both, the factor for yellow seeds somehow dominated the factor for green seeds (the *recessive* factor). Mendel argued that the recessive factor remains unaltered and was just as likely to be passed on to future generations as the dominant factor. This enabled Mendel to explain, among other things, why one quarter of the offspring from two heterozygous plants which each produce yellow seeds will produce green seeds.

Mendel provided a theory of inheritance, but he also provided a model for how further studies could be conducted. Mendel had achieved his success by focusing on discontinuous traits, by ensuring that his original stock was breeding true (for example, self-fertilized tall plants would only produce more tall plants), by conducting large-scale and carefully controlled cross-breeding experiments. Mendel's work

[3] Kuhn (1962, 36).

was an incentive for others to see whether further species and traits would instantiate Mendel's ratios, and so provided guidance on future work.

There is a lot more to Mendel's work than I have summarized here, but hopefully we've seen enough to illustrate some important Kuhnian ideas. Mendel focused others' attention on certain aspects of the experiences under investigation. Studies of biological inheritance could start with a range of different questions and approaches which might lead investigators in quite different directions. The Mendelian paradigm focused attention on discontinuous traits, breeding experiments and the proportions of subsequent generations that exhibited those traits. Mendel's conclusions weren't adopted because the evidence Mendel gathered was unassailable. He certainly hadn't established that his ideas held universally. His influential work was based on the study of just one species! Nevertheless, Mendel's work was (eventually) recognized as interesting and novel, and appeared promising for purposes of addressing issues surrounding biological inheritance.

Subsequent researchers attempted to replicate the studies, expand the theory to incorporate more species and traits, and develop the ideas in more detail and depth. Among those who shared a sense of admiration for Mendel's project, there was thus a level of agreement concerning how questions of inheritance might profitably be addressed and what kinds of evidence and arguments were worth developing. The detailed work of classical geneticists, furthermore, which from the outside might appear rather trivial, paved the way for portentous developments in the life sciences of the twentieth century. As I've portrayed the science here, Mendelian genetics fits the Kuhnian notion of a paradigm quite nicely. This is a good thing – the intent was to help illustrate Kuhn's view about what science is. However, we should be cautious with such examples. The history of Mendelian genetics is far more interesting and nuanced than the brief sketch I've provided. The extent to which Mendelian genetics instantiates the concept of a Kuhnian paradigm is a question for serious historical and conceptual study, and the same is true for any period and concentration of scientific investigation.

There are further important concepts in Kuhn's account, but before reviewing these let's pause to reflect on some important lessons from our discussion of pre-paradigm science and normal science. First, the key difference between these periods, for Kuhn, is understood principally by attending to qualities of *communities* of scientists. Within pre-paradigm science, there is insufficient consensus between working scientists for progress to take place, but in normal science community-wide agreement is apparent, and progress is possible. A striking feature of Kuhn's claim is that we can't properly understand the nature of the sciences by attending solely to the role of evidence and observation, and the kinds of arguments and inferences that scientists make. According to Kuhn, we must also recognize sciences as social activities, conducted by groups of people who influence and interact with one another. This is a suggestion that gives rise to many additional questions, some of which we will raise in the next chapter.

Second, a significant cost of ignoring the historical and social dimensions of scientific inquiry, for Kuhn, is that we are likely to overlook the fact that methods, assumptions and scientific interests are all subject to change, as are the standards by which scientific arguments are evaluated. We are likely, that's to say, to impose our own ideas on past debates and thus be guilty of a kind of historiographical bias. Kuhn denies there is a universal and unchanging scientific method that we should be motivated to articulate and defend. The sciences have been more dynamic along more dimensions that we typically presume. Kuhn's concept of a paradigm helps sell these points. Scientists who work within a given paradigm share a set of ideas that are more specific than a commitment to some universal scientific method and a desire to gather evidence that supports the core theoretical assumptions. These more specific ideas won't all be shared by scientists who work within other paradigms. The core components of a paradigm, furthermore, are not challenged within normal science; by definition normal science involves fidelity to certain core assumptions and practices, and their *protection* from falsification as normal science unfolds. This conclusion stands in stark contrast to traditional empiricist views, Popper, for example, according to which any scientific idea is vulnerable to falsification if the right data are gathered. Normal science doesn't include deliberation over fundamental issues.

Third, although a particular and notable scientific achievement sets off a period of normal science, Kuhn emphasizes that the seminal work never answers all questions that investigators were previously engaged with, and it immediately confronts reasonable objections, phenomena it can't explain and observations that conflict with the theory. New ideas are pursued, at least in part, because they are judged a promising avenue for exploring scientific interests. Finally, Kuhn seems impressed by the extent to which normal science focuses the attention of practising scientists towards incredibly narrow and esoteric details. This is an aspect of the sciences that is very hard for outsiders to appreciate. In the final chapter of the book, we will spend some time discussing genetically modified food and scientists' ability to take genes from one species and transfer them into another. We won't describe the science in any detail, but even a cursory glance reveals that the development of this technology required enormously wide-ranging theoretical understanding and the development of methods designed to manipulate chromosomes, plasmids, phages, restriction enzymes, ligase, recombinant DNA, gel electrophoresis and a bacterium that introduces its DNA into plants. In the process of these research efforts, thousands of restriction enzymes have been identified; countless genes, chromosomes and genomes have been sequenced; and even more plants have been cultivated. Quite simply, an extraordinary amount of labour, time and effort is invested in every aspect of the technology, pursued by multiple laboratories around the world over the course of several decades, resulting in tens of thousands of technical scientific papers appearing in hundreds of scientific journals, having been reviewed and critiqued by hundreds of editors and reviewers. Whatever we make of Kuhn's concept of normal science, we should strive not to lose sight of the magnitude of modern scientific research, and the breadth and depth of research that lies behind many scientific results.

Scientific Revolutions – Kuhn's scientific paradigms are eventually overturned and replaced. In describing how this occurs, Kuhn starts by noting that normal science itself gives rise to anomalies. Observations are made that conflict with expectations based on extant theories; data are gathered, but can't be explained from within the paradigm. Some anomalies will be resolved. Those that aren't won't cause immediate concern or calls to abandon the paradigm. Hope lives on that these anomalies will be removed without violence to the paradigm, once someone provides the pivotal insight. However, the longer anomalies persist, the more troubling they become. Some anomalies might take on particular significance for other reasons. As they accrue and their resolution becomes more urgent, scientists start to lose confidence in the existing paradigm. At some point, suggests Kuhn, the anomalies become sufficiently worrisome that normal science enters a *crisis state*. The principles and assumptions that were previously immune to challenge and debate now look less certain. *Something* about the community's core, working assumptions must be mistaken, for otherwise these anomalies would yield to scientists' efforts to explain them. With respect to at least some assumptions, scientists must be mistaken. A state of crisis is one of two criteria necessary for a scientific revolution.

As scientists begin to suspect that their paradigm lacks the means to resolve its anomalies, confidence in that paradigm diminishes. A scientific revolution, however, requires not just a sense of crisis surrounding an old paradigm, but the appearance of a new exemplar that possesses the right blend of success and promise. This is an important point and worth reflecting upon further. An empiricist attitude towards the sciences might lead us to expect that, once anomalies have accrued, a scientific community is obliged to abandon the affected theoretical framework and begin working on a new approach. Kuhn certainly rejects this as a good description of how the sciences actually proceed, and there are reasons to wonder whether science *should* abandon a framework that can't explain some of the phenomena that we think it should.

First, every theory ever proposed has faced challenges. Philosopher of science, Imre Lakatos, would later say that, in an important sense, all scientific theories 'are born refuted and die refuted'.[4] Second, the problems with one approach may afford little insight into what alternatives are available or sensible. Paradigms provide crucial guidance concerning how phenomena should be investigated, how data should be gathered and what questions it is sensible to ask. When released from such guidelines, it may be quite unclear how we are to proceed with investigations. It might be prudent, therefore, to continue with what remains the most promising of available techniques and to use this as a platform for ongoing research. As alternatives are developed, we

[4] Lakatos (1978, 5).

would hope that these are afforded attention, but only when an alternative emerges that solves extant problems and exhibits genuine promise might it become reasonable for the scientific community to start transferring allegiances.

Revolutions require crises and rivals. Once the revolution is complete, normal science can resume, but now science operates under new management – a different paradigm prevails and hence different ideas on how science should be done. For Kuhn, the kinds of change that take place within normal science are very different from the more dramatic lurches that are associated with revolutionary change. As we've emphasized, paradigms don't just include ideas about what kinds of things exist, but also how scientific inquiry should be conducted and how the results of that inquiry should be evaluated. Whether a particular puzzle solution is acceptable or not, whether data have been appropriately gathered or not, whether data have been appropriately analysed or not, and so on – issues such as these are settled by guidelines and principles that are *internal* to the paradigm. Paradigms include the rules for how science is conducted and how its achievements are judged. A scientific revolution, however, by changing the paradigm, changes the rules by which scientific ideas are evaluated. There is certainly historical evidence of changes in methods and standards of evaluation. An important and interesting case concerns the reception of Newton's work on universal gravitation.

Newton's theories appeared at a time when the sciences were dominated by a mechanistic philosophy; good scientific explanations were supposed to appeal to push and pull forces between particles that came into direct contact with one another. Newton's suggestion that massive objects attract other massive objects via a gravitational force, despite being very far removed from those objects, sounded to many of Newton's contemporaries like a return to discredited forms of scientific explanation. Attributing to objects the capacity to influence at a distance the motion of other objects wasn't considered good science. Convincing other scientists of Newton's law of universal gravitation required, in part, revising existing ideas on what qualified as an acceptable form of scientific explanation.

Since revolutionary science threatens the defining components of a paradigm, including standards of evaluation, then evaluating a scientific revolution must seemingly appeal either to a more general set of criteria, suitable for evaluating theories from distinct paradigms, or we must concede that revolutions unfold in the absence of any guidelines. Kuhn's view is that there are no helpful rules which can guide practitioners within revolutionary science. Considerations such as the simplicity of available theories, or their precision, or scope might play a role when it comes to choosing between theories from different paradigms, but these notions are vague and their application to particular cases unlikely to yield definite verdicts. Kuhn concedes, even embraces, the remaining option: there are no rules that govern theory choice within revolutionary science.

If scientific revolutions are not guided by rules, then what does cause a community of scientists to abandon one set of ideas in favour of another? In response to this

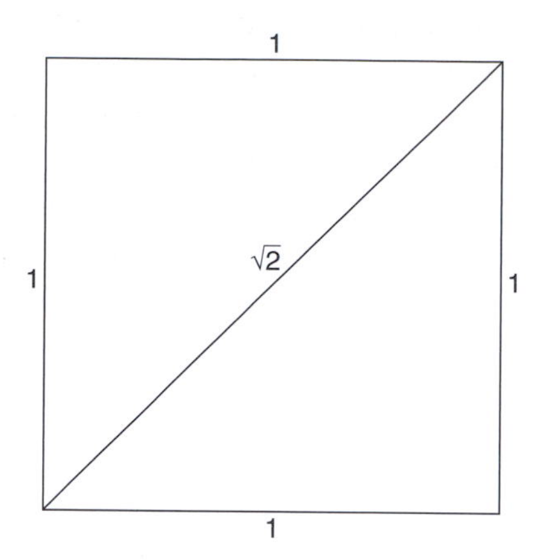

Figure 3.1 The diagonal of a square is incommensurable with the sides, because they cannot be expressed as a ratio of whole numbers.

important question, Kuhn draws an analogy between scientific revolutions and gestalt images. The latter are familiar. A simple line drawing looks like a rabbit, but if you shift your attention slightly it can appear instead like a duck. (The old woman/young woman illusion in Chapter 2 is another example.) When your experience changes from observing a duck to observing a rabbit, however, it is not a change in the image that induces that change. The world doesn't change, but your experiences of the world do. Kuhn thinks that something analogous occurs within scientific revolutions. The arrival of a new paradigm provides a new way of looking at the world, a new way of pursuing scientific research, and new ways of evaluating the kinds of conclusions that are generated. Kuhn's explanation here is very thin on details, as he himself admitted. What provoked far more discussion and attention than Kuhn's suggestive remarks about gestalt switches, however, were his negative views on how scientific revolutions *aren't* resolved. Perhaps the most widely discussed aspect of all Kuhn's views is his suggestion that distinct scientific paradigms are *incommensurable*, a conclusion which has important consequences for our notions of scientific rationality and progress.

To say that two things are incommensurable is to assert that they lack a common measure. The term has its origins in ancient Greek mathematics. If we consider the diagonal and side of a square, then their lengths cannot be expressed as the ratio of two integers (i.e., whole numbers). If we stipulate that the sides measure one unit length, for example, then the length of the diagonal can only be expressed as what is now called an irrational number (Figure 3.1). There is no unit or measure that allows us to describe the diagonal as having length x units and the sides as having length y units, where x and y are whole numbers. The diagonal and side of the square are, in this sense, incommensurable.

Kuhn argues that scientific theories from distinct paradigms are incommensurable. In important respects, the theories lack a common measure. Kuhn's thesis is subject to competing interpretations, and his views on incommensurability seem to have undergone some change. On more modest interpretations, Kuhn is arguing that instances of theory comparison are more complicated than were previously assumed, that we lack justification for supposing that more recent theories are somehow *nearer the truth* than the predecessors, that there is no measure of *truthlikeness* or *proximity to the truth* and that the outcome of scientific revolutions isn't predetermined by available evidence. More radical interpretations, which Kuhn later attempted to distance himself from, supposed that incommensurability establishes that instances of theory choice are not rational. On the latter interpretation, if theories cannot be compared, then there could never be rational grounds for concluding that one theory is *better* than another. In this chapter, we'll focus on the more modest interpretation, saving discussion of more radical conclusions for Chapter 4. Thus far, however, we have said little of *why* Kuhn regards competing paradigms as incommensurable. There are several reasons.

First, Kuhn claims that 'the proponents of competing paradigms will often disagree about the list of problems that any candidate for paradigm must resolve.'[5] Suppose a scientific paradigm asks and answers certain kinds of question, yet is ultimately replaced by a paradigm that dismisses some of those questions as meaningless, confused or irrelevant. Insofar as these paradigms have distinct goals, comparative assessments will be more complicated. Second, Kuhn defended the idea that observations are theory laden, along the lines discussed in the previous chapter. A common empiricist assumption, prior to Hanson's and Kuhn's work, was that the plausibility of observation statements is not affected by an individual's theoretical commitments. Kuhn was not alone in suspecting that this might not be true, that attempts to describe what we experience are affected by our expectations, which may be influenced by our theoretical commitments. Thus Kuhn suspected that scientists from distinct paradigms might not even agree on the observational data, which could certainly obfuscate efforts to compare theories from distinct paradigms.

Third, Kuhn draws some radical conclusions from the fact that the *language* of scientific discourse changes. For example, although Newton and Einstein both used the term *mass*, the meaning of this term changed. If two scientific communities employ the same terms to mean different things, then (according to an influential thesis in philosophy of language called *semantic holism*) that difference will have ripple effects throughout the entire language. If you embed the contested term within a sentence, the old and new sentences will mean different things, depending on who is uttering them. The meaning that is intended by the speaker will depart from the

[5] Kuhn (1962, 148).

meaning that is inferred by the listener if they attach different meanings to just one of the terms within the sentence. According to Kuhn, working within different paradigms hinders, and perhaps effectively prevents, meaningful communication.

Finally, beyond working on different problems, disagreeing about observations and using technical terms in different ways, any changes in standards of assessment will also contribute towards incommensurability. Confronted with two theories, and two standards of evaluation, we might find that one theory looks better from the perspective of one set of standards and the other theory appears better from the perspective of the other standards. How can scientists reach reasonable judgements if they have different ideas about how to judge? As we saw, Kuhn doubts that there are more abstract criteria that could help guide decisions.

Even if we credit Kuhn only with endorsing the more modest version of incommensurability, it remains a thesis of great interest and importance. The modest thesis denies that there exists a formula or procedure for resolving all anomalies that paradigms might confront, or for resolving debates between competing paradigms. For example, Kuhn doubts that so-called *crucial experiments* are as decisive as they might appear: confronted with competing hypotheses or theories, one might suppose that the debate will be definitively settled once we develop a test, or crucial experiment, for which the hypotheses predict distinct outcomes. On historical grounds, Kuhn doubts that this is how disputes have generally been resolved. (We will consider further reasons to be suspicious of crucial experiments in the next chapter.) The modest thesis further implies that we should be more loath to start describing individual scientists as either rational or irrational because we might lack justification for attaching these attributes, and that resolutions to some scientific debates might be more open-ended than we're inclined to suppose. None of these consequences entails that the resolution of scientific revolutions is irrational, but we will postpone further discussion of this issue until the next chapter.

Scientific change within normal science is for Kuhn fundamentally different from change as it occurs across paradigms. Within the former, as data are gathered, conclusions and hypotheses are retracted, modified, corrected and replaced. The merits of these changes can be evaluated according to the standards of the paradigm. The assumptions that can't be challenged, and don't change, define the paradigm. Changing these involves a paradigm shift. Any alteration to the core components of a paradigm represents something other than normal science. Kuhn suggests that, as he uses the term, scientific revolutions are not limited to the major accomplishments of Copernicus, Lavoisier, Darwin, Einstein, and so on. A scientific revolution might affect only a small community of scientists and go largely unnoticed outside that community. However, if a change to the core of their existing paradigm occurs, usually preceded by a state of crisis among them, then this qualifies as a revolution. The consequences of revolutionary science for those affected are truly revolutionary in Kuhn's eyes. He describes scientists from different paradigms as living in different worlds, unable to appreciate or understand the perspective of other paradigms.[6]

Conclusions

Kuhn's small book is packed with big ideas that provoked a huge reaction. We've now sketched Kuhn's central theses. It's beyond the scope of the chapter to consider many of the worthwhile responses to Kuhn's work, but several conclusions deserve further emphasis, because they help inform our evaluations of issues that will become important in the final part of the book.

First, Kuhn does us a service by insisting that it is unhelpfully simplistic to suppose that all scientists employ the same basic method and hence that all scientific change can be understood as the accumulation or revision of conclusions and results, which follow from combining that (unchanging) method with observations and data. We have emphasized the absence of a universal method in earlier chapters, but Kuhn did more than most to promote the idea, claiming that it lacks historical plausibility. We betray the versatility of scientific investigation if we impose a single scientific method on all sciences then reject particular scientific theories on the basis that they don't employ that method. Yet this is the mistake that creationists make when, for example, they distinguish historical sciences from observational sciences, then argue that only the latter furnishes us with reliable conclusions, and hence that we can justifiably reject anything that falls outside their definition of an observational science. Reasonable evaluations of scientific ideas require engaging the goals, methods and types of evidence of that particular discipline.

Second, Kuhn's historical work reveals that the role of anomalies and persistent puzzles is quite unlike what we might expect on the basis of naïve empiricist account. For example, Kuhn describes how scientific paradigms are initially adopted by scientific communities despite being incomplete, unable to account for certain kinds of phenomena and perhaps less suited than competing approaches to resolving some problems. Paradigms are attractive for the promise they hold for future success, for their success with respect to a certain kind of problem and perhaps for further reasons that vary from case to case and are therefore more difficult to articulate. Kuhn describes the most important work that's conducted within normal science (as well as the work that is most prevalent within normal science) as directed towards articulating the paradigm. It would be a mistake to suppose that scientific theories are presented to the world complete, unambiguous and without defect. Furthermore, incompleteness, ambiguity and deficiency will accompany a paradigm until it is replaced; these are not reasons for disappointment, however, but stimuli for new

[6] The idea of scientists living in different worlds sounds very strange. We might suspect that Kuhn must be talking metaphorically, but this is not easy to align with some passages in Kuhn. Some commentators interpret Kuhn as a kind of constructivist, hence sympathetic to the idea that reality is constructed through our scientific theories. Kuhn subsequently tried to distance himself from these constructivist interpretations.

research. (And of course these qualities of a paradigm are quite consistent with that paradigm also providing very convincing explanations for a wide variety of phenomena.) When sceptics mine scientific journals for signs that a given theory is due for replacement, or was initially pursued for less than incontrovertible reasons, then we shouldn't be surprised that they uncover anomalies, shortcomings, ambiguities and contradictions. All research programs include these features. Those who malign the theories of mainstream science shouldn't expect us to abandon these theories until alternatives emerge that are at least as promising.

A related point concerns those critics of mainstream sciences who accuse scientists of dogmatically persisting with certain ideas despite contradictory evidence, and of accommodating the problematic observations by introducing additional hypotheses that lack independent motivation or support. The dogmatism is supposedly evident from the fact that scientists haven't changed their mind about the core issue, but have instead offered responses that may sound speculative rather than well confirmed. But again, sometimes a theory is sufficiently well supported by available evidence that it is far more reasonable to regard anomalies as arising from as-yet-unidentified factors, rather than evidence that the entire theory is bankrupt. In particular cases, we might worry that scientific communities have paid insufficient attention to alternative perspectives, but we have no reason to suppose that these problems are systemic.

Kuhn's central idea, however, is that of a scientific paradigm, and here we must be more careful about what we can learn from Kuhn, where Kuhn himself may have gone wrong, and where others may have misinterpreted what Kuhn intended. In defence of the most general idea, there does appear some historical plausibility to the idea that scientific communities tend to coalesce around certain practices, assumptions and theories, and hence something like the concept of a Kuhnian paradigm does appear to play a role in the sciences. Furthermore, some of the key features that Kuhn attributes to paradigms do very plausibly play an important role in the emergence, maturation and solidifying of many scientific disciplines: extensive, influential research programs have been launched by a single, seminal, novel approach to some phenomenon; the attention of practising scientists within many disciplines and sub-disciplines is thereby directed towards particular problems and techniques, and certain assumptions and attitudes become prevalent; the professionalization of scientific disciplines and the education of new generations of scientists is an important aspect of any discipline's success; and certain ideas may not seem acceptable targets for significant challenge or critique. There are, however, disciplines and research programs that instantiate the concept of a Kuhnian paradigm far less well, so we should be cautious about how much weight we attach to Kuhn's account.

Beyond doubts surrounding its scope, there are other aspects of Kuhn's account that seem over-stated. Kuhn suggests that scientists working within normal science possess an almost overwhelming commitment to the core tenets of a scientific paradigm, hence a very strong reluctance to countenance certain kinds of change.

But, first, it seems likely that scientists are often quite capable of working within a paradigm, and thus assuming for the purposes of their research that its core assumptions are correct and unchallengeable, while simultaneously being quite comfortable with the idea that evidence could emerge that would warrant changing the paradigm in some respect.[7] Second, even if some scientists are dogmatic in their attitude towards core assumptions, a community of scientists is likely to include a range of attitudes towards different methods, assumptions, results and so on. Some might be dogmatic and others far less so – a possibility that has important consequences for our attitude towards the rationality of scientific communities as opposed to individual thinkers. Kuhn's theory of normal science, and the degree of commitment he attributes to working scientists, also raises for him the question of why paradigms would ever be overturned. Kuhn's own solution is to propose the appearance of crisis states as a condition for scientific revolution, but the historical evidence for crisis states is not strong. With hindsight, we might recognize groups of scientists as struggling to solve anomalies because they are labouring under certain false assumptions, but that's not evidence of scientists' worrying that their core assumptions are mistaken. Kuhn himself admits that the explicit recognition of crises by scientists is rare.

Perhaps Kuhn was guilty of exaggerating certain claims. We can retain much of his theory while diluting some of the core ideas. It is clear that scientists are often *more* reluctant to amend or replace some ideas than others, but we needn't suppose that they are incapable of being persuaded to abandon even highly entrenched ideas. Similarly, it is likely that education is hugely formative for individual careers. Of course, this will make it harder to appreciate alternative approaches and perspectives – but not impossible. In general, I suspect that the distinction between normal and revolutionary science is often fuzzier than Kuhn implies. As noted, Kuhn includes within the work of normal science the articulation of a paradigm, but elsewhere suggests that any change to a paradigm qualifies as a scientific revolution.[8] In many cases, however, it may be hard to say whether a theory is being articulated or revised. It was Kuhn's triumph to recognize that we cannot shoehorn all scientific change into a single overarching methodology, but we should be cautious about following Kuhn's lead and supposing that there are two sharply divided kinds of change.

These criticisms have consequences for our understanding of Kuhnian incommensurability. Even the modest version of that thesis, as we noted above, deserves serious reflection. If Kuhn was arguing for something stronger, then his success is far less clear. In Chapter 1, we considered in some detail the idea that observations are theory laden. The best we can say here in support of Kuhn's thesis of incommensurability, I think, is that gathering reliable data can be more difficult than we might suppose, but this will at worst complicate, not prevent, meaningful theory comparison. Standards of evaluation

[7] Godfrey-Smith (2003, 84).
[8] See Bird (2001, 41) for evidence that Kuhn sometimes seems to regard any change to a paradigm as revolutionary.

do change, but this need not thwart our efforts to reach reasonable judgements about competing approaches either. In some cases, it is quite uncontroversial that some methods are *better* than others. Once we understand the placebo effect and experimenter's bias, it is immediately apparent why double-blind trials are preferable. There is no obvious reason to deny that scientists typically adopt alternative standards because they have good reasons for doing so. Kuhn's concerns about changing languages tap into some very difficult problems in the philosophy of language, but we may not need to resolve these to satisfy ourselves that something must be wrong about Kuhn's view. As Peter Godfrey-Smith notes, 'If incommensurability of meanings is real, as Kuhn says, then it should be visible in the history of science. So those who study the history of science should be able to find many examples of the usual signs of failed communication–confusion, correction, a sense of failure to make contact.'[9] Godfrey-Smith's impression is that such examples are very rare. Each of these factors makes some scientific changes more difficult to evaluate; they don't obviously preclude the possibility of reasonable assessments. Of course, it would be nice to offer a more positive conclusion, about the possibility of reasonable comparisons between competing scientific paradigms, an idea that we will return to in the next chapter.

Kuhn's influence is substantial, and his book more than repays a very careful reading. Communities of scientists do share certain assumptions, both about the world and the best ways to investigate it. Both types of assumption can change, so attitudes towards how scientific inquiry should be pursued will sometimes change. Theories are not embraced because they answer all our questions. They immediately face challenges, but they also give us cause for optimism that interesting progress on new fronts might be achieved. Scientific revolutions don't occur as a result of anomalies alone, and the outcomes of revolutions may be more contingent, and hence less predictable, than we're inclined to suppose. However, there is nothing in Kuhn to suggest that any given scientific idea can't be overturned, if we're presented with the right arguments and data. Scientists can be reluctant to give up certain ideas, but this needn't be attributed to stubbornness. Ideas are prized because there exists a substantial body of evidence that supports them, because they play a central role in a coherent and successful research program and because collectively these far outweigh a few unexplained observations. Only a naïve empiricist attitude towards science would suppose that scientific theories can be uncontroversially falsified by a single experiment.

Discussion questions

1. Should scientists abandon a theory as soon as it confronts anomalies? Why or why not?
2. Is it sensible to draw on historical considerations when reflecting on the nature of scientific research?

[9] Godfrey-Smith (2003, 92).

3. Normal science happens within a Kuhnian paradigm, and revolutionary science occurs when one paradigm is replaced. Do you think it is useful to make a sharp distinction between these two types of scientific inquiry?
4. Is scientific expertise something that can be learned solely by reading books, or is it something that requires hands-on experience, whether it be conducting experiments, gathering data, being part of a research team, teaching others about the discipline and so on? If the latter, are these reasons to have more or less confidence in scientific expertise?
5. Is it reasonable to suppose that all scientists trained in a particular field are largely incapable of evaluating the profound challenges to their basic scientific assumptions, due to the psychological hold that paradigms have over them?
6. Do you think that scientific research programs are typically pursued because they appear a promising means of investigating some domain? How does this affect your view of the sciences?
7. Would it be a problem for our attitude towards science if we discovered that some scientists have stubbornly persisted with ideas that others regard as clearly discredited? Why or why not?
8. How might paradigms aid scientific progress, and how might they hinder it?
9. If some scientists are more inclined to question even core assumptions of a paradigm, might this help explain the emergence of scientific revolutions more satisfactorily than Kuhn's appeal to crisis states?

Suggested reading

Kuhn's most famous book is well worth reading cover to cover, at least a few times (Kuhn 1962). Hacking's introduction to the fiftieth anniversary edition is also very worthwhile. There is an incredible amount of secondary literature on Kuhn. Hoyningen-Huene (1993) is exceptional. Bird (2001) is also very good. All introductory texts to the philosophy of science include at least one chapter on Kuhn's theory of science, each offering slightly different interpretations. Worthwhile examples include Chalmers (2013), Hacking (1983), Ladyman (2001) and Godfrey-Smith (2003).

4
Sciences as historically and socially situated

One of Kuhn's greatest influences on the philosophy of science – an influence quite independent of the merits of his own theory of science – was his use of historical and sociological considerations for purposes of addressing traditional, philosophical questions. Kuhn wasn't the first to think about either the history or sociology of science, but these were nascent disciplines, and their relevance to philosophy was largely ignored. Over the last few decades, the history and sociology of the sciences have each flourished as academic subjects. Simultaneously, philosophy of science has become far more sensitive to the possibility that understanding the sciences requires paying attention to their histories, social contexts, the psychology of working scientists and the social organization of scientific communities. To a significant degree the sciences are now recognized as responses to changing human and social interests, pursued by people of varying personality, celebrity, socio-economic upbringing, race and gender, who work within scientific communities that compete for funding and influence, which generate conclusions that may be unpalatable or inconvenient to certain groups in society. A theory of the sciences that regards them as simply an abstract set of methods and assumptions can produce, at best, an impoverished account.

There is undoubtedly much we can all learn from historical, sociological and psychological perspectives on the sciences, but such perspectives are also often associated with radical and profound challenges to our understanding of what the sciences can achieve. These sceptical arguments threaten scientific authority and the idea that sciences represent our most reliable methods for understanding the world. In this chapter, we'll introduce the arguments. We'll see that their significance for evaluating created controversies is not straightforward. Those who create controversies are typically more concerned with securing for themselves a degree of scientific legitimacy, an objective which isn't well served by arguments that challenge the possibility of scientific authority. Nevertheless, certain critics of mainstream sciences continue to advance versions of these sceptical arguments, and thus it is beneficial to spend some time reviewing them. Towards the end of the chapter, we will survey several further projects that take very seriously the idea that sciences are historically situated, social activities. In some cases, these accounts recognize ways in which sciences can be improved, but they maintain a generally optimistic view of scientific success and the possibility of achieving scientific progress.

In Chapter 3, we distinguished stronger from more modest interpretations of Kuhn's thesis of incommensurability. The modest version supposes that there is no clear procedure by which we can adjudicate between the merits of theories from distinct paradigms, and that, for various reasons, comparing those theories might be significantly more challenging than we perhaps typically suppose. For example, of many scientific debates we can't rely on a crucial experiment to unequivocally determine which theory is correct; hence there may be nothing straightforwardly irrational about continuing to explore either. The stronger thesis, however, suggests that there is no rational basis for judging one paradigm better than another, and hence theory choice cannot be rational. The modest thesis challenges the suggestion that evidence and reasoning are sufficient to uniquely determine how scientific debates and disagreements should be resolved. It leaves open the possibility that more than one resolution to a debate is rational, justifiable and reasonable. The strong incommensurability thesis supposes that evidence and argument are so impotent, relative to the resolution of scientific debates, that they can do little, or perhaps nothing, to constrain the possible outcomes of those debates. To understand *why* certain scientific theories become popular, and why certain scientific choices are made, the strong thesis implies that we cannot look towards the available evidence, and so must look elsewhere for explanations. Some readers have interpreted Kuhn as suggesting that scientific revolutions are settled not through evidence and argument but instead via the politics of scientific communities and broader social factors. We'll ignore the question of whether Kuhn is most charitably read as an advocate for the strong or modest version of incommensurability. Even if Kuhn wasn't advancing the strong thesis, others have subsequently defended something very much like it. It is to these thinkers that we now turn.

Scepticism and the sociology of science

Sociological studies of science predate Kuhn, but in the 1970s there emerged more ambitious and more radical projects. These thinkers sought to replace philosophy of science, rather than augment it. They strived to explain why scientists reached certain conclusions about the world, not in terms of evidence, testing and analysis, but in terms of social factors, like influence, community-specific standards and personal ambition, as well as wider social tastes, desires, expectations and so on. Subsequent sociological studies of science seemed to push even more radical theses, suggesting that scientists are actually responsible for *constructing* knowledge, facts and reality. The development and proliferation of these programs and ideas involved not just sociologists of science, but historians and feminist scholars, and, to a lesser extent, philosophers, cultural anthropologists and political scientists. Collectively, these attitudes and approaches towards science became most widely known as Science and Technology Studies (STS), and by the 1990s there arose media interest in the so-called Science Wars, within which many members of the STS community were regarded as occupying one side of this impassioned debate.

The Science Wars are hard to adjudicate, in part because the camps on both sides include a variety of different views and ideas. In broad outline, what appeared to be at stake were scientific objectivity, the rationality of the sciences and the possibility of scientific progress. Defenders of traditional attitudes towards the sciences regarded their opponents as hopelessly confused about how sciences function, and of peddling theories of the sciences that were incoherent. Those who allied themselves with the more radical elements of STS insisted that historical, sociological and philosophical arguments compel the conclusion that sciences are just one way of investigating the world and that they can lay no legitimate claim to being the best way. We won't survey much of the expansive literature that surrounds these various arguments, but two very influential ideas are worth reflecting upon, as well as a more general, epistemological concept that seems to underpin much recent scientific scepticism. The first big idea to consider is that scientific facts, scientific knowledge and even reality are all *socially constructed*.

There is now a long list of things that, it has been argued, are socially constructed. Ian Hacking identifies books of the form *The Social Construction of . . .*, or *Constructing . . .*, for Authorship, Brotherhood, and the Child Viewer of Television, all the way through the alphabet to Youth Homelessness and Zulu Nationalism.[1] Only the letter X is missing from Hacking's list; in 2001 sociologists from Uppsala University, Stockholm, published *The Social Construction of Xenophobia and Other Isms*.[2] As Hacking describes, however, it is not always clear what is intended when something is described as having been socially constructed. Some things are clearly social constructions. The borders that separate nations, states and provinces are the results of rather arbitrary agreements among groups of people to divide the land a certain way. There is no deep, ontological reason behind these decisions, nor for the decisions to group the days into sets of seven, giving rise to the concept of a week, divide hours into sixty equal parts or measure distances, volumes and areas according to those units that have now become standard.

Many of the concepts and ideas we understand and use regularly might appropriately be described as social constructions. Hacking's analysis unearths further interesting ways in which social constructions can profoundly shape our experiences. For example, whether someone qualifies for certain medical treatments, welfare assistance, educational scholarship, political refugee status and so on, will be decided

[1] Hacking (2000). [2] ftp://ftp.cordis.lu/pub/improving/docs/ser_racism_burns.pdf.

by rules and criteria that committees have agreed to. Whether given individuals just about satisfy these criteria, or marginally miss them, could make an enormous difference to their future. To the extent that such criteria are chosen arbitrarily, individuals are deeply affected by social constructions.

From these modest observations, we can conclude that much of the world, as we encounter it, is one of rules, criteria, boundaries and conventions, many of which have been selected by society, perhaps for good reasons, but without their reflecting any deep metaphysical truth about reality. However, there are thinkers who go beyond these commentaries, suggesting that *everything* is a social construction, including that which is described by our best scientific theories. Importantly, the thought here is not simply that our *theories* of gravity, viruses, atomic structure and neurology are socially constructed nor that our *ideas* about gravity, viruses, and so on are socially constructed – these would be less controversial. The implication is instead that gravity and viruses and atoms can *only* be understood as social constructions. Roughly speaking, the worry is that since we cannot access a reality that's independent of our experiences, then the only meaningful things we can say about reality must be constructed from our ideas and theories. As critics have pointed out, however, these kinds of constructivism confuse the construction of ideas and theories that are intended to *represent* aspects of reality, with the construction of reality itself.[3]

Figuring out how we represent features of the world, how we can evaluate the quality of our representations, how we improve our representations and so on are difficult, important questions. That these issues are challenging, however, does not make them incoherent, does not collapse the distinction between representations of reality and reality itself and thus fails to threaten scientific objectivity in the manner some social constructivists seem to imply. The ways in which scientific results become accepted *facts* within scientific communities raise worthwhile questions, and careful ethnomethodological approaches might provide useful insights.[4] Nevertheless, the merits of such studies again provide no reason to collapse the above distinction between represent*ations* and what's represent*ed*.

Some sociologists of the sciences are constructivists. An earlier generation are better described as relativists. These latter thinkers hoped to explain why scientific theories are accepted and how scientific debates are resolved, by appealing entirely to social factors. The sciences are one important mechanism by which information is generated. We might suppose that scientific communities embrace particular results because the evidence advanced in favour of them is compelling and that individual members of those communities are well positioned to evaluate and recognize the force of the arguments. Rational argument is not the only means by which beliefs

[3] For a helpful perspective on these issues, see Kitcher (2003, 24–28, 51–53, 59–62).

[4] Latour and Woolgar (1979) is a very influential book within sociology of science that adopts what anthropologists call an ethnomethodological approach. Latour and Woolgar observed the daily activity of a scientific laboratory, from which they sought to understand how scientific facts are *constructed*.

might become popular within some community, however. Propaganda, education, superstition and familiarity – each affects social groups and the spread of ideas. Of particular beliefs we might hope to identify the reasons that a community came to hold them. Advocates for the *strong program* in the sociology of scientific knowledge were curious about the reasons that scientific communities accept certain scientific conclusions, but rather than attempt to explain the acceptance in terms of the evidence available, or the merits of the methods being utilized, those who defended the strong program suggested that the popularity of scientific ideas should be understood in terms of social factors. Central to their methodology was the *symmetry principle*.

According to the symmetry principle we should seek to provide the same kinds of explanation for a community's adoption of some set of beliefs, regardless of our own evaluation of those beliefs' truth or falsity, and regardless of our own evaluations of the rationality of the methods or practices that produced them. The same kind of explanation should be sought for purposes of understanding how scientists came to believe that water is H_2O as would be sought if we're concerned to understand why an ancient civilization came to believe in reincarnation. We would violate the symmetry principle if we looked to explain the former by appealing to the fact that water *is* H_2O, or by suggesting that the methods that produced this conclusion were a reliable means of discovering chemical composition. By denying that we can justifiably prefer some methods and conclusions over others, advocates for the strong program were committed to a kind of *relativism* about scientific knowledge: sciences adopt certain standards of evidence, justification and rationality, but other communities adopt alternative standards. Advocates denied that some practices are more rational than others, or more likely to yield true conclusions. The strong program thus implied that there are no better or worse descriptions of the world, no truths or falsehoods among our beliefs; instead there are only beliefs that are regarded as true, according to some ways of investigating the world, or are considered more or less plausible, according to certain ways of evaluating claims.

David Bloor's *Knowledge and Social Imagery* was the first book-length introduction to the strong program. Writing fifteen years after its publication, Bloor noted that the book, and the strong program, had 'won few friends and many enemies'.[5] There were admittedly variations among the ideas developed within the strong program, but also a shared affinity for a strong form of Kuhnian incommensurability and a shared reluctance to explain the acquisition of community belief in any terms other than social forces. The program was heavily criticized for never properly explaining *how* social factors have affected theory choice or development. Its plausibility often seemed to rest on the *possibility* that certain ideas would grow in popularity if social and political factors aligned in their favour, without providing evidence that they had achieved popularity for such reasons. A useful perspective on these issues is provided

[5] Bloor (1991, 163).

by introducing a further important assumption for those who were sympathetic to the strong program. The possibility that scientific theories are *underdetermined* by available evidence can help shed much light on debates within recent philosophy of science.

Underdetermination

To illustrate the concept of underdetermination let's suppose you're attempting a crossword puzzle.[6] One of the clues is *European capital city*. The answer has six letters. You've already entered into the puzzle answers to other clues and these suggest that the fourth letter is L and the final letter N.

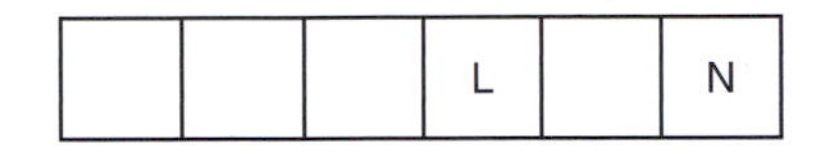

Figure 4.1

You're about to write in *BERLIN* when it occurs to you that *DUBLIN* also satisfies all the criteria you've been given. The information you have is consistent with more than one answer. In other words, the information that's available doesn't determine a unique correct answer; the information *underdetermines* the solution to the question we've been asked. Very often, the information that's available to us is insufficient to warrant conclusions with certainty, because we recognize that our information is consistent with more than one explanation. If I arrive for a department meeting and discover that no-one else is present, I might wonder whether the time or venue has been changed, whether the meeting has been cancelled or whether I made a mistake about the meeting time or location. Until I gather more information, all these possibilities are consistent with available evidence. The possibility that our scientific theories are underdetermined by available evidence has spawned a significant literature and plays an important role for those who regard sociology of science as a suitable replacement for the philosophy of science.

The names most often associated with the underdetermination of scientific theories are those of Pierre Duhem and W.O. Quine. Duhem was a French physicist, and a historian and philosopher of science, who worked in the late nineteenth and early twentieth centuries. One of Duhem's arguments for underdetermination asks us to consider a scientific hypothesis, a prediction derived from that hypothesis and a test of the prediction which produces an outcome that is inconsistent with the prediction. As we saw in our discussion of Popper's falsificationism, a failed prediction doesn't straightforwardly falsify our target hypothesis. Perhaps there was a problem with the

[6] As with many philosophical concepts, not everyone understands underdetermination in quite the same way. Stanford (2013) is a thorough and very good introduction to the topic.

instruments being used or an unknown confounding variable. The derivation may have involved sophisticated mathematical techniques which introduce the possibility of error in the prediction itself. Duhem argued that this *holistic* feature of hypothesis testing, whereby testing a hypothesis invariably requires making a host of further assumptions, undermines the idea that science can advance by constructing *crucial experiments*. It is an example of underdetermination because available evidence doesn't uniquely determine how we should account for the discrepancy between derivation and experimental outcome.

Duhem's discussion was limited to the field of physics. In the middle of the twentieth century, Quine argued that underdetermination applied more generally. Quine's theory of knowledge was that *all* our beliefs are tested against experience whenever any of our beliefs are tested. Quine describes how:

> The totality of our so-called knowledge or beliefs, from the most casual matters of geography and history to the profoundest laws of atomic physics or even of pure mathematics and logic, is a man-made fabric which impinges on experience only along the edges. . . . A conflict with experience at the periphery occasions readjustments in the interior of the field. But the total field is so underdetermined by its boundary conditions, experience, that there is much latitude of choice as to what statements to re-evaluate in the light of any single contrary experience.[7]

Quine also writes that 'any statement can be held true come what may if we make drastic enough adjustments elsewhere in the system'.[8] For Quine, disagreements between our experiences and the predictions we derive on the basis of our beliefs about the world can be overcome in many different ways. Even beliefs about logic, mathematics and the definitions of very familiar terms – beliefs that appear to us thoroughly incontrovertible – are, in an important sense, revisable. Quine admits that there are many beliefs that we will feel extremely loath to revise or modify, but for Quine this reflects psychological facts about us rather than something distinctive about the beliefs themselves. On the basis of available evidence, some inferences seem far more sensible than others, but Quine denies that these inference rules possess the kind of privileged status we might feel inclined to attribute to them. Such rules are revisable, if we are confronted with the right body of experience.

Quine's theory of knowledge has profound implications for any theory of rationality, but its threat to scientific rationality has attracted most attention. If additional empirical data is unable to settle scientific disputes, because there are always many ways of reconciling failed predictions with pre-existing theoretical commitments, then something other than observation and evidence must be functioning to settle those disputes. As noted, Quine acknowledges that some reconciliations will appear far more reasonable than others. My speculation that the department meeting has been moved seems more plausible than the idea that my colleagues are extra-terrestrial visitors who have now returned to their home planet. So what makes some

[7] Quine (1951, 39). [8] Ibid, 40.

underdetermined solutions to outstanding puzzles appear more reasonable than others? Quine suggests that we prefer those that are simpler, and that require least disruption to previous beliefs, to those that are nearer the edges of that corpus of beliefs that confronts experience. However, these criteria for theory choice are, according to Quine, only of pragmatic value.

Scientific choices and disputes should be explained in terms of human psychology, according to Quine. Sociologists of scientific knowledge are more wont to explain such outcomes in terms of social and political factors, the ambitions, loyalties and biases of practising scientists and the communities they comprise, their pursuit of influence, funding, status and so on. Such explanations are attractive to a certain style of thinking: we might understand very little about why the germ theory of disease, genetic theory of inheritance or atomic theory of matter became widely accepted; it thus introduces some cognitive ease to suppose that the same general explanation applies in all cases, an explanation which appeals to human and social traits that are familiar to all of us. However, as noted above, an important criticism of socio-political explanations for scientific change is that advocates have failed to demonstrate in any detail how social factors have influenced the actual content of scientific theories. As we will see shortly, the ambitious sociological accounts exemplified by the strong program are further weakened when we reflect more closely on the thesis of underdetermination itself.

Before we offer some critical remarks on underdetermination, it is worth pausing to reflect on the idea of *contingency* within scientific development. As was apparent in Kuhn's work, the absence of any general procedure by which all scientific disputes are settled implies that scientific disputes *could* have been settled in other ways and hence the history of particular disciplines could have unfolded along quite distinct lines. According to some historians, for example, embryology played a surprisingly small role within the development of the big ideas of twentieth century biology, suggesting that if embryologists had achieved more influence then the history of biological thought might have looked quite different. Sociologists of scientific knowledge have embraced the idea that scientific advancement is contingent, arguing that distinct social, economic and political conditions could have dramatically altered the outcome of scientific debates.

Historical contingency is inconsistent with the idea that the sciences *had* to proceed as they actually proceeded. It is less clear whether it threatens scientific authority, however, and hence it is unclear how troubled we should be by the possibility that history might have been different. First, much historical contingency is likely attributable to individuals and communities being motivated to understand particular phenomena, or solve particular problems, where it is conceivable that other phenomena could have attracted their attention instead. However, contingencies of this variety allow that, once a particular ambition is articulated, the range of adequate solutions will be very heavily constrained. Having set out to understand why the first half of the twentieth century experienced a rise in reported cases of lung cancer, it was perhaps extremely likely, given some further plausible assumptions, that the cause would be traced to the rise in cigarette use. Contingency can be admitted, because we might have pursued questions other than those we actually pursued, but without threatening scientific authority, because we're supposing that once a particular question is articulated then a single, most reasonable answer will become apparent.

Second, contingency only threatens the possibility of scientific *progress* if we suppose that as a scientific discipline advances then there is always one unique best approach to unsolved problems and unexplored questions. But this may not always be the case. Suppose there are competing research programs available, each reasonably well confirmed and each demonstrating a similar degree of future promise. In such circumstances, it might not always be unreasonable to pursue just one of these options and investigate how much progress it can help us achieve, where of course there is nothing to stop us later returning to a previously abandoned approach. Alternative histories might converge on very similar scientific results; thus again we see a form of contingency that presents no clear threat to scientific objectivity or progress.

Finally, we should be careful not to suppose that it is irrational to reject a particular theory only when its foundations have been incontrovertibly discredited, or the core assumptions of the preferred approach been established with certitude. Within any scientific debate, evidence and arguments accumulate. Working within one research program might become increasingly irrational even if there is no precise moment at which continuing to investigate that program becomes unreasonable. The fact that scientific disciplines could have evolved along very different trajectories is not itself a threat to scientific authority. However, the possibility that we are currently ignoring scientific theories which are as good as, or better than, our own does deserve further discussion, bringing us back to the thesis of underdetermination.

Kyle Stanford provides a helpful perspective on the issue of scientific underdetermination, a perspective that involves distinguishing *global* underdetermination from *local* underdetermination.[9] Global underdetermination occurs when we have competing scientific theories that make all the same predictions about whatever it is that we can measure, observe or discover. If there are such alternative theories, which describe the world differently but are indistinguishable from one another through any of their empirical consequences, then confidence in our own theories might seem misplaced. We can have no reason to prefer our own theories, it would appear, over their empirically indistinguishable rivals. Quine's thesis entails that there are such rivals. As Stanford argues, however, examples of global underdetermination are reminiscent of many radical, sceptical hypotheses, and they are no more consequential.

In the early seventeenth century, René Descartes considered the possibility that an evil and powerful demon could be responsible for all Descartes's sensory experiences

[9] Stanford (2006, 12–16) and Stanford (2013).

and memories, and hence that almost everything that Descartes took himself to know was based on hallucinations. A modern variant of this radically sceptical hypothesis suggests that I might be a brain in a vat, wired to a huge super-computer that is sending electrical signals into my brain and thereby simulating for me the impressions that I have a body, am drinking coffee, while sitting at my desk, listening to music, and glancing out of the window. I have many beliefs about my history, my surroundings, and my physical body, but all these beliefs are rendered open to doubt by the possibility that a demon or super-computer is creating these experiences for me. Radical scepticism is familiar to philosophy. Stanford's observation is that the evil demon and brain-in-a-vat scenario are cases of underdetermination: in the same way that a unique correct solution to my crossword puzzle was underdetermined by the information available, similarly, whether I am being deceived by a demon, or a super-computer, or am in fact sitting at my desk, drinking coffee, and glancing out of the window are all underdetermined by my sensory experiences.

There *could* be scientific theories, quite unlike those currently being entertained yet predicting all the same consequences, and I *could* be a brain in a vat. Each are consistent with my experience. Typically, of course, the possibility of deceitful evil demons doesn't affect our reasoning and formation of beliefs. Perhaps there are lessons to be learned by reflecting on radical scepticism, concerning very general questions about the nature of knowledge and justification, for example. Insofar as underdetermination focuses our attention towards *scientific* ideas, it can motivate good questions about scientific methods, arguments and inferences, their assumptions and justification. If we're looking for inferences that appear reasonable, given assumptions that we're willing to grant, we could learn a great deal. If we develop the habit of doubting all assumptions, then we enter the realm of radical scepticism.

In Chapter 4, we recognized a similar problem with our positive perspective on the problem of induction: if we can offer positive arguments for scientific conclusions only by first admitting certain further assumptions, then demands to justify those further assumptions will arise. Such demands, however, must either continue indefinitely, terminate with assumptions that are self-evident and hence require no further justification, or circle back on claims that have already been made and hence beg the question. None of these three options appear satisfying, but the trilemma is not one we should hope to solve.[10] Nevertheless, conceding that inductive scepticism and Cartesian scepticism are insoluble challenges doesn't entail that we can't provide compelling arguments, given assumptions that most people will find reasonable.

When establishing whether a sample of sodium is pure, checking the details of a mathematical derivation, observing the behaviour of a predator in its natural habitat or

[10] Within mainstream epistemology, these three (seemingly exhaustive) options make up Agrippa's trilemma, the basis of an important sceptical argument. Of course, even if an uncontroversial solution to the trilemma looks unlikely, there might be much value in considering the relative strengths and weaknesses of the three alternatives. For a helpful introduction, see Klein (2008).

discovering the spectral lines of a distant star, we don't worry about evil demons or the possibility that we're a brain in a vat, nor should we worry about global underdetermination. Exactly what assumptions are being made in each of these cases is challenging to articulate and evaluate, but this shouldn't discourage from us insisting on identifiable and persuasive reasons before relinquishing plausible conclusions. Quinean underdetermination falls short of providing such reasons. Quine is so permissive – in the kinds of beliefs he allows us to adjust – that his thesis is no more relevant to our evaluations of particular claims than any radical, sceptical hypothesis.

Reflecting again on Quine's suggestion, that the totality of our beliefs is 'so underdetermined by its boundary conditions, experience, that there is much latitude of choice as to what statements to re-evaluate in the light of any single contrary experience', we might push for reasons to suppose there is *much* latitude, rather than a tiny bit. No amount of evidence can definitively compel a conclusion that extends beyond the data themselves. However, once we begin to augment our data with rules of inference that we agree are reasonable, and with objectives that we value, then we will generate reasons to prefer some conclusions over others. I don't know what kinds of adjustment to our corpus of beliefs would be required in order to deny that viruses cause disease, that organisms are made of cells or that planets orbit the Sun. It certainly may, however, involve a conspiracy of truly Cartesian proportions. The degree of latitude that Quine describes within the choices for belief revision might be as extensive as Quine suggests only if we are willing to entertain a great many, competing, radically sceptical theories.

Local underdetermination, by contrast, is described by Stanford as involving a small number of claims within a more general scientific theory, where these claims are not settled by available evidence, and which thereby represent a kind of wiggle room within the broader set of beliefs that we assume are not underdetermined. The evidence we either have available, or are capable of gathering, might strongly recommend a broad range of beliefs about the world, but residual questions could remain unanswered.

The most careful, persistent, skilled biographer of Abraham Lincoln might fail to confirm what Abraham Lincoln ate for breakfast on his ninth birthday. Perhaps knowing everything that there is to know about the world as we find it today would offer no good evidence about the contents of Lincoln's childhood breakfast. However, this particular failing does not mean that we can't learn an enormous amount about Lincoln's life and times. It is reasonable to have enormous confidence in much of his biography, even if some questions are unanswerable. Science might be like this. From our vantage there might be unanswerable questions, questions for which multiple answers are all equally sensible, given available evidence. However, as Stanford suggests, we have no obvious means of estimating either the number of these questions or their significance. Thus, there appears nothing problematic about assuming evidence will settle many of the questions we raise, while striving to ensure that we attend to alternative hypotheses that fit our data equally well. When cases of local underdetermination are uncovered we withhold assent. So long as communities of scientists respond to instances of local underdetermination appropriately, the latter don't appear to pose an interesting objection to scientific authority.

Finding distinctive, serious challenges to the reliability of scientific theories in neither global nor local forms of underdetermination, Stanford introduces a third variety. *Transient* underdetermination involves those occasions when theories or hypotheses generate alternative predictions, but where we have thus far been unable to gather sufficient evidence to adjudicate between them. As Stanford is aware, transient underdetermination also offers no immediate, substantive threat to scientific objectivity. If we discover that we have competing hypotheses, each similarly well confirmed by available data but generating distinct predictions concerning possible future observations, then the prudent course is to explore each hypothesis further in the hopes that subsequent data will help settle the debate. Stanford senses a deeper problem lurking in the vicinity, and offers interesting considerations in support of his conclusion, although not everyone is convinced that the problem is serious.[11]

The possibility that our scientific theories are underdetermined by evidence has had a profound impact on twentieth-century discussions of scientific objectivity and rationality. A prevailing concern has been that evidence and norms of scientific rationality can determine neither theory choice nor the appropriate revision to a system of beliefs that yields a false prediction. The strong program was influenced by these arguments. An important motivation for pursuing sociological explanations for the endorsement of scientific conclusions was the belief that theories are so underdetermined by available evidence that the latter cannot account for why given theories were preferred. Yet on closer inspection it's hard to see whether there are distinctive challenges arising from underdetermination-based arguments. If the concern is that our theories *might* be in error because we might be unwittingly ignoring alternative theories that are just as good as our own, then we should be neither surprised nor troubled (and should remember that we might be brains in vats): unsurprised because we always recognized that science is fallible; untroubled because the possibility of error is consistent with having very compelling evidence for given conclusions. If the concern is instead that we might have asked different questions of our experiences and the data we gather, and thereby have identified distinct goals for our scientific inquiries, then we should concede the contingency of scientific research but note that this concession does nothing to undermine the merits of the answers we

[11] In outline, Stanford is concerned that scientists have a habit of defending certain types of scientific conclusion on the basis that no alternative explanation is even conceivable, only for future generations to advance explanations that are not just different but judged superior. Insofar as sciences rely on promoting theories by eliminating all serious alternatives, Stanford worries that we have historical evidence that sciences are not reliable. For objections to Stanford's argument, see Magnus (2010), Godfrey-Smith (2008) and Chakravartty (2008).

have produced in response to the goals we've in fact pursued. If the concern is that there are no universal rules of rationality, at least none with sufficient specificity to direct scientific inquiry across all disciplines, then we should consider whether the absence of such rules is a significant handicap: in particular circumstances our sciences are capable of advancing very compelling arguments for their conclusions. If the concern is that alternative scientific methods, rules and techniques could have satisfied the goals and ambitions that we pursue, but generated very different theories in the process, then we should seek reasons beyond mere possibility before taking the concern seriously. It might be that many of the alternatives for reconciling experience with theory would either change very little about our core beliefs, or would involve changing our values, goals and objectives so significantly that they are not true competitors to our theories but radically different ways of living in the world.

None of these interpretations provide the profound challenge to scientific objectivity that many seem to have supposed, at least not with further argumentation. On at least some of these interpretations, we discover that there are important issues that deserve further attention; furthermore, admittedly, arguing that underdetermination *isn't* the radical challenge to scientific authority that is often suggested doesn't provide us a *defence* of scientific authority, something we will return to later in the chapter. Most importantly, however, dismissing particular scientific results requires a better argument than the mere possibility that there might be a better explanation out there somewhere.

Scepticism emerges from the history of science

If we understand underdetermination as the rather weak claim that *it is possible* that there are radically different and wholly unconceived scientific theories which would satisfy curiosity as well as our own theories and fulfil the practical ambitions we place on our sciences, while respecting the restrictions we place on how evidence should be gathered, the limitations on what we can achieve due to technological, financial and cognitive limitations, then what becomes critical to our evaluation of the significance of underdetermination is how seriously we need to regard this possibility. Are such possibilities speculative, flights of fancy, or is the existence of such alternatives likely? In reflecting on this question, history may have an important role to play.

If we suppose the history of the sciences has been one of pursuing broadly similar goals, then we might expect scientists to achieve broadly similar results. However, even a cursory glance through the annals of history reveals that the histories of the sciences are ones of refinement and change, and sometimes profound change. Among many other things, history is a graveyard of abandoned scientific methods, theories, ideas and attitudes. The humoral theory of medicine, phlogiston theory of combustion, caloric theory of heat and optical aether theory, are among many theories that at one time attracted significant support but were subsequently overturned. Perhaps the historical record provides some evidence that we should take underdetermination seriously; perhaps the historical record provides evidence that there very likely are radically distinct alternatives to today's theories which fulfil the same criteria. Regardless of their relationship to underdetermination, historically based sceptical arguments deserve our attention.

The history of the sciences includes ample evidence of change, which might lead us to wonder why we should have any confidence in today's scientific announcements. Some current scientific investigations will no doubt result in the modification of existing scientific theories and understanding. Recognizing that our scientific theories could require some degree of revision is unsettling. Should we attach a warning label to every scientific result, advising that this is based on current understanding and subject to changes that we can anticipate in neither detail nor scope? Should all scientific claims be taken as equally likely to require modification, or are some less vulnerable than others? Are there ways to identify less vulnerable conclusions? Climate change sceptics remind us that in the 1970s some scientists were predicting a new ice age. Those sceptics invite us to infer that climate science is highly unstable, liable to change, and that we shouldn't place much emphasis on today's conclusions because tomorrow's might be different.[12] At the very least, it appears not straightforward how we should balance the dynamic nature of the scientific process, and thereby the complication that today's scientific ideas will change, against the authoritative status that we think today's scientific conclusions deserve and that in certain contexts we find ourselves depending upon.

The significance of the observation that past scientific theories have been successful, yet ultimately replaced by new and fundamentally different theories, has played a particularly crucial role for philosophers of science and for the debate between scientific realists and antirealists. At stake within this debate is what attitude we should adopt towards those scientific conclusions and assertions that describe objects which we cannot observe directly. Physicists describe a world populated by protons, leptons, quarks and other exotica. Likewise chemistry and biology describe objects that we can't observe, except perhaps with sophisticated apparatus. Scientific realists see no reason to deny that we can achieve substantive knowledge and understanding, even of objects that are too small for us to see with the naked eye. Antirealists, by contrast, admit that unobservable entities may be very useful ideas, since they enable us to predict and control at least certain kinds of events and circumstances. Nevertheless, for the antirealist the utility of these ideas isn't a good reason to suppose that they reflect the nature of a world independent of our experiences.

The most influential argument in favour of scientific realism draws attention to scientific success. Many realists are sufficiently impressed with the achievements of

[12] It bears noting that while some climate scientists were predicting global cooling, during the 1970s, there were more climate scientists predicting global warming. As well as a questionable argument based on history, climate change sceptics are, in this case, also guilty of cherry-picking, a general strategy that we will return to in Chapter 7.

modern science that they suppose it would be a miracle if our best scientific theories weren't at least approximately true.[13] How could our theories be so successful, the realist contests, if those theories grossly misrepresented what they purport to describe? An influential antirealist response to this challenge draws attention to history. Explaining how theories can be successful, and yet profoundly mistaken, might be no easy task. However, many discredited scientific theories appeared successful but are, by current lights, not even rough approximations to their modern replacement. The fact that today's theories are wonderfully successful can't be a reliable indicator that those theories are approximately true. History is full of examples of successful theories which now appear hopelessly misguided.[14]

Realists have had a lot to say in response to this historical argument. Many have questioned whether past theories were as successful as antirealists sometimes imply. Perhaps the humoral theory of medicine, catastrophist theories of geology and Darwinian theory of pangenesis were successful by some standards; however, as we've noted, standards of evaluation change, and the theories that have been replaced don't always meet the more rigorous standards that we now employ. The rejection of past scientific theories doesn't straightforwardly compromise the merits of our own theories if the latter are being held to higher standards of evaluation.

Changing standards complicate our hopes of drawing reasonable inferences from the history of science about present or future changes, and they are not the only confounding factor. Ludwig Fahrbach, for example, draws attention to the exponential growth of the sciences, both in terms of the number of scientists who are active and the number of articles being published.[15] Fahrbach estimates that 80 per cent of all scientific work has occurred since 1950. There is also more interdisciplinary work than ever before, as evidenced by the emergence of new interdisciplinary disciplines like biophysics, palaeoanthropology, quantum chemistry and so on. The size of today's science, relative to that of generations past, the rise in collaborative research and the more exacting standards that are demanded each undercut efforts to reliably predict anything about the future of the sciences, based on the past. In many respects the present is unlike the past, and some of these differences provide reasons to suppose that today's sciences are far better than those they replaced, and hence the mere fact that scientists have been wrong before is a poor reason for supposing that they're wrong about particular claims they advance today.

A second form of realist response reminds us that history records more than a sequence of scientific revolutions. Those who draw attention to history, for purposes of challenging current science, emphasize the radical changes that have occurred, but history also records cases of significant continuity and stability. Ian Hacking describes how the laboratory sciences 'lead to an extraordinary amount of rather permanent

[13] See Psillos (1999) for an important introduction to, and defence of, scientific realism.
[14] The canonical presentation of this objection to realism is Laudan (1981). [15] Fahrbach (2011).

knowledge, devices and practice'.[16] In many disciplines, we seem able to identify sequences of theories that gradually correct for the deficiencies of earlier approaches, draw on technological, conceptual and empirical developments and thereby produce models and theories that better agree with observations, and achieve far greater accuracy and scope. If science is to make progress, then change is entirely necessary. If progress involves more than the accumulation of facts and, in particular, may involve learning more about which methods work, how methods can be improved, the limitations of our apparatus, how we interact with the world and so on, then it is not surprising that the resulting history reflects no clear patterns. If our *goals* for scientific research are prone to some revision or augmentation those histories will become even harder to unpick.

The general point here is that history doesn't obviously weaken the possibility that we are now in possession of scientific theories that provide far better representations of their targets and that consequently we possess substantive knowledge of at least some aspects of the natural and social worlds. Of course this is not to say that history *recommends* such positive attitudes. However, if we remember that histories of the sciences involve not just change, but also stability, continuity and refinement in response to identifiable deficiencies, improved methods and explanatory gaps, then it becomes much harder to see why previous mistakes are a good reason to doubt current convictions.

Neither the underdetermination of scientific theories by evidence nor the radical discontinuities apparent from the histories of the sciences provide straightforward reasons for seriously doubting scientific authority, although each serves as an important reminder of fallibility. Similar arguments and considerations have been alluded to by those who seek to undermine confidence in scientific theories that are judged inconvenient or distasteful, but in one important respect such objections extend beyond what creators of controversy typically aspire to. Thus, even if someone was to object that we have underestimated the force of these sceptical arguments (and I am happy to admit that there are many nuances and details we have ignored), there remains an important difference between emphasizing the fallibility of science and mounting a convincing criticism of particular scientific conclusions or assumptions.

When the cigarette industry sought to undermine public confidence in the medical profession's claim that cigarette smoking increases the chances of developing lung cancer, their target was not *all* science but a very particular scientific result. Insofar as spokesmen for industries and other special interest groups appeal to the fallibility of the sciences to undermine confidence in particular theories, they provide no better reason to doubt the *targets* of their attack than any other scientific claim. What matters isn't that we *might* be wrong, but how likely it is that we *are* wrong and in what ways. Conceding that current scientific understanding will be altered tells very little, if

[16] Hacking (1992).

anything, about the future of particular scientific claims, and hence the fact that scientists have been wrong before is a weak argument for doubting specific, modern scientific results.

Evaluating scientific communities

Thus far in the chapter we have been largely concerned with sceptical arguments, but there are undoubtedly important and more optimistic insights to be gained about the sciences from adopting social and historical perspectives. Explanations for scientific change that draw on social factors might enrich and complement our understanding of particular developments, but to do this they needn't *replace* histories that describe the available evidence and argumentation. Miriam Solomon argues that much traditional thought on the nature of the sciences promotes a strongly *individualistic* attitude. Sciences – according to such accounts – are about gathering data and utilizing suitable methods to generate justified beliefs: those who gather data with sufficient care and apply methods judiciously will generate reliable conclusions. Scientific communities are rational or irrational, from these perspectives, to the extent that each of their members is rational or irrational. Individualistic approaches largely ignore the role of trust, collaboration and competition. Furthermore, by supposing that individual scientists can successfully and usually suppress personal or social goals, such accounts underestimate the force of these factors. (In Chapter 5, we will review several important cognitive biases. Our focus won't be on scientists' vulnerability to such biases in particular, but it would be naïve to suppose that individual scientists are not affected. Of some such biases, furthermore, empirical evidence does suggest that scientists commit these errors.)

Evidence that individual scientists are prejudiced, biased and influenced by social factors might easily be taken as evidence that the sciences are not rational. However, Solomon suggests that the fault may lie not in our assumption that science is rational, but in supposing that scientific rationality is a product of the rationality of individual scientists. Her contention is that scientific success and progress is better analysed at the level of scientific communities, rather than of the individuals that comprise those communities. This need not be taken as implying that individual scientists are all hopelessly irrational, only that the *greatest* strength of modern sciences may lie in their social organization rather than the skill and ingenuity of their members.

To illustrate the basic idea, suppose that all scientists were equally and strongly reluctant to adopt new methods and approaches, entertain new theories and ideas or trust the reliability of surprising data and discoveries. Under such conditions, sciences would become stagnant and uninteresting. However, if all scientists were similarly and sufficiently disposed towards trying novel methods, developing speculative ideas and rejecting prevailing wisdom, then scientific communities would likely become heavily splintered, effective communication would become difficult and well-confirmed ideas would be forgotten. Balancing the value of successful, entrenched ideas against the promise of novel ones does not permit easy solutions. Plausibly, however, the sciences are more likely to achieve a good balance through a mix of attitudes among working scientists, rather than instilling within every individual some ideal balance, assuming there is such a thing. Insofar as science makes progress, relative to certain objectives, this might be most plausibly attributed to the fact that there are conservative scientists, who give up on old ideas only very reluctantly, and reformist scientists, who are more inclined to pursue novelty. Clearly, there is significant work needed to articulate these ideas with precision, but hopefully the example at least helps motivate the idea that social organization might be relevant to scientific success.

Solomon is broadly concerned with the idea that scientific rationality and success are best evaluated at the level of scientific communities, rather than at the level of individual members who comprise that community, while remaining alert to the fact that a wide variety of factors can influence individual scientists. These factors include an individual's desire for credit, competitive spirit and reluctance to admit arguments that threaten one's own conclusions. Solomon wasn't the first to consider the role of such factors within the sciences. David Hull, for example, argues that scientists are motivated by the desire for recognition, and consequently individual scientists strive to contribute something *new*.[17] New work might fill lacunae, introduce alternative and competing hypotheses, correct prior mistakes or seek novel perspectives on traditional questions.

One intriguing consequence of Hull's observation is that, insofar as the sciences reward those who correct previous errors or who introduce well-supported new ideas, the community of scientists protects itself against dogmatism and stagnation. To be clear, no-one is arguing that such communities are entirely immune to these vices, only that communities which value self-correction, novelty and rigorous standards for those who claim such achievements, thereby do more to protect themselves than communities that don't.

Another interesting project from social epistemology – the study of social aspects of knowledge – was initially explored by Philip Kitcher and concerns the distribution of cognitive labour.[18] Suppose we have a shared desire to investigate certain phenomena and several viable techniques that each appears promising. As a society we could recommend that all relevantly trained scientists pursue that approach which seems most promising. However, in some circumstances it might be more sensible to distribute labour across these various approaches more equitably, perhaps with more or less resources being directed in particular directions according, in part, to how promising they appear.

[17] Hull (1988). [18] Kitcher (1993, ch. 8).

Kitcher is particularly interested in what kind of reward system would motivate individual workers to distribute across the competing programs in the optimum way. For example, if there are larger individual rewards associated with being part of a group that achieves success through a less popular, less promising avenue, then this provides some motivation for workers to pursue those less popular options. The details surrounding Kitcher's argument have not convinced everyone, but the spirit of the project is very appealing. Reflecting on questions like the distribution of labour appears a very worthwhile pursuit for purposes of considering whether some communities are more or less likely to achieve identifiable goals.

Returning to Miriam Solomon's work, she directs particular attention towards the concepts of scientific consensus and scientific dissent. By the late nineteenth century, biologists had accepted that species change over time, but there was disagreement concerning the mechanisms of change. Darwin had presented and argued for his own theory of natural selection, but other biologists were considering other explanations for the evolution of species. Within geology, for much of the first half of the twentieth century, the suggestion that continents move relative to one another was not widely accepted although it did enjoy minor support. At around the same time, Mendelian genetics was experiencing considerable ascendancy, but this came at the expense of various alternatives. Solomon discusses these three examples of scientific disagreement, or dissent, among others, as well as instances of scientific consensus. Her concern is to develop a framework with which we can start evaluating whether instances of dissent and consent are reasonable.

Within that framework Solomon distinguishes between the roles of empirical and non-empirical factors for scientific theory choice. The former provide reasons to prefer particular scientific theories because such factors measure empirical successes, although Solomon notes that particular successes might be more or less available to different scientists, and more or less salient; hence different scientists will be more or less heavily influenced by them. Non-empirical factors include an individual scientist's pride, her preference for radical alternatives versus entrenched concepts, deference to authority, peer pressure and so on. Of cases of scientific dissent Solomon hopes to evaluate the relative empirical successes of competing approaches, as well as the distribution of non-empirical factors, and their relative dominance across some period. The simultaneous pursuit of multiple approaches, and hence the continuation of a particular scientific debate, might be appropriate and justified, as we've seen. Theories that enjoy more success should receive comparatively more attention, for Solomon, but she further suggests that achieving equitable (i.e., proportional) distribution of cognitive effort also requires an even distribution of non-empirical factors.

One of Solomon's examples is Mendelian genetics. According to Solomon, while Mendelian genetics and embryologists each experienced empirical success, the former received disproportionately favourable attention. The reason, for Solomon, was that more non-empirical factors were pulling in its favour. Early advocates for Mendelian genetics had more political influence than the embryologists, their projects received generous funding from plant and animal breeders who hoped to benefit financially from these techniques, and the United States university system was well suited to accommodate the new discipline. These factors, suggests Solomon, rather than its empirical credentials, provide more convincing explanations for the relative dominance of genetics during the early decades of the twentieth century. More generally, Solomon's view is that an uneven distribution of non-empirical factors will likely result in an inequitable distribution of effort, which would be non-ideal with respect to scientific progress. In several further cases Solomon concludes that the relative dominance of particular approaches was appropriate and can be traced to an even distribution of non-empirical factors.

Solomon's framework for evaluating instances of scientific dissent can also be utilized to evaluate cases of scientific consensus. As Solomon observes, consensus is a special case of dissent, when all empirical success within some domain belongs to one theory and where, as a result, non-empirical decision factors should slowly disappear. Solomon thus agrees with Kuhn that the mere presence of anomalies or shortcomings does not in itself bring a particular scientific consensus into question. Justified dissent from the consensus requires at least some empirical successes that count in favour of the alternative stance. Insofar as those who create scientific controversies are inclined to place enormous emphasis on the inadequacies of one theory, rather than the strengths of an alternative, they are, from Solomon's perspective, engaged in activity that couldn't legitimize genuine scientific dissent.

The social empiricism espoused by Solomon is not a feminist critique of sciences which warrants mention only because there have been important feminist philosophies of the sciences that are often concerned with sciences' social organization. One important and ongoing concern of such philosophies is that the sciences continue to perpetuate and enforce existing gender stereotypes. Rosalind Franklin conducted scientific research in the early 1950s that played an important role in the discovery of the structure of DNA. Unfortunately, Franklin's contributions were not given the credit she deserved, and her treatment by male scientists was extremely lamentable. Prejudice and inequality remain problems for the sciences, just as they do within society at large. Several feminist critiques of sciences argue that the enormous obstacles for women entering and succeeding within the sciences are not only a matter of social injustice but actually hinder scientific progress.

A nice example of scientific progress being achieved through the introduction of more women into a field is that of primatology and, in particular, the study of the social and sexual behaviour of chimps, gorillas and other nonhuman primates. Theories of such behaviour had, in the first half of the twentieth century, attributed to females a very passive role. Social behaviour, including sexual behaviour, was thought to be controlled by male members of the group with alpha males playing a particularly important role. In the 1960s, several female researchers entered the field of primatology and began to gain some prominence. Jayne Goodall became well known for her work with chimpanzees, and Diane Fossey with gorillas. More female researchers entered the field in the 1970s. It was during this time that the prevailing scientific theories of primate social behaviour began to change. Observations revealed a more active role for female primates. Females manipulated male behaviour in subtle and important ways, and exerted control over who they mated with.

It seems very likely that our understanding of primate social behaviour benefited from the influx of female researchers. Helen Longino argues that the sciences generally would profit not just from better gender equality but greater diversity of working scientists in terms of socioeconomic upbringing, ethnicity and so on.[19] Whether such advantages would appear across all disciplines is an open question, but alternative perspectives on scientific issues are important; encouraging recruitment and funding policies that would promote these goals is entirely apposite.

A similar conclusion emerges from case studies that reveal how information and knowledge from certain practices have been imprudently overlooked by scientists. Sociologists, historians and philosophers have all discovered examples where scientific studies could have performed better, because experts who fell outside what is regarded as the traditional scientific community were ignored despite possessing pertinent information. For example, a nuclear power station disaster in 1986 created a problem for parts of the United Kingdom after weather systems transported radioactive material to the UK from the disaster site in Chernobyl, within the former Soviet Union. Government scientists in the UK were charged with evaluating the risks to consumers who ate meat from animals that had grazed on the affected geographical regions and, if necessary, with introducing policies to minimize those risks. The scientists have since been criticized for introducing methods that weren't sensible and, in particular, for ignoring the local farming community who protested the methods and attempted to explain why they were impractical.[20]

This and similar case studies are important reminders that we still have more to learn about best scientific practices, that mistakes continue to be made and that many groups have important contributions to make in resolving socially significant issues. These are aspects of scientific investigation that would be overlooked if no-one was attending both to the internal social organization of scientific communities and the ways in which scientists interact with other groups within society. However, we should be careful with more general conclusions we might draw from such examples. Discovering useful methods for evaluating scientific research, and finding examples of study that could have been conducted more efficiently, is quite consistent with

[19] Longino (1990).

[20] This example is described in more detail in Collins and Pinch (2002). Barker and Kitcher (2013) describe further examples.

discovering that much scientific research is remarkably well designed with respect to the described goals, remarkably successful at helping us reach accurate and useful conclusions and where we can appreciate these conclusions for ourselves when presented with the evidence.[21]

Sciences and values

One factor that Miriam Solomon describes with regard to the political success of Mendelian genetics was the corporate interest of plant and animal breeders. Large-scale businesses have often been disparaged for their influence over certain scientific fields. Pharmaceutical companies are criticized for their influence over product testing within the health professions. It is likewise suggested that multinational bioagricultural companies lack sufficient oversight concerning the testing and safety of their products. When faced with such worries, it is tempting to suggest that the problem is not with the *sciences*, but with political interference or lack of adequate independent monitoring. The response presupposes that we can fully distinguish sciences from *values*. Heather Douglas has argued that we cannot.

Douglas challenges the *value-free ideal*, which she describes as the idea 'that social, ethical, and political values should have no influence over the reasoning of scientists, and that scientists should proceed in their work with as little concern as possible for such values'.[22] Douglas argues that the sciences cannot sensibly aspire to the ideal, but also *should* not aspire to the ideal. Scientists play a hugely important role as advisers to public policy formation, and this practically ensures that societal concerns and values will enter scientific dialogue. We want to avoid suffering and other harms, but we also value privacy, freedoms, innovations, justice, biodiversity, the environment and so on. That we fund scientific research demonstrates that the pursuit of certain kinds of knowledge is itself valued by society. That we insist on scientific research meeting certain ethical standards demonstrates that scientific work, thankfully, is restricted by further, important values. That which we value affects both what we research and how we research. Sciences are not driven solely by the interests and values that are internal to its disciplines, but also respond to general societal and ethical values; hence they cannot be value free.

The value-free ideal appears unattainable, but Douglas also argues that it is undesirable and that scientists have a moral responsibility to carefully consider the potential social and ethical consequences of publishing and publicizing work that might be in error. Of course Douglas is fully aware that the possibility of error accompanies all empirical work, but some inaccuracies will have more costly

[21] In several respects the case involving the UK farmers is unlike much scientific research. For example, there was urgency to complete the work expediently, despite several very significant complexities in gathering reliable evidence, evaluating it and introducing strategies to reduce risks.

[22] Douglas (2009, 1).

consequences than others. Suppose that we have empirical evidence suggesting that a cheap and simple policy change will likely improve certain safety standards and that thousands of lives could potentially be saved. The evidence is uncertain, as all scientific evidence is, but advisers must decide whether to recommend the policy change. Importantly, there is in this case an important asymmetry between recommending and failing to recommend the policy change. If the policy is recommended, but subsequent investigation reveals that the policy change will be ineffective, a small financial cost has been incurred. Failing to recommend the change, however, and later discovering that the policy would have saved many lives, incurs a much greater cost. For Douglas the *consequences* of error can't be ignored, and in matters of public policy evaluating those consequences will seldom be value free.

That values will and should play important roles within the sciences is, for Douglas, importantly distinct from the possibility that values function as reasons for *accepting* certain claims. Although we might have values that create a desire to believe certain claims, those values do not provide *support* for the claims. Cigarettes cause lung cancer, whether you like it or not. Relinquishing the value-free ideal, and hence admitting that values play an essential role in science, does not imply that we can justifiably reject scientific conclusions simply because we don't like them. Thus for Douglas there are legitimate and illegitimate roles for values within the sciences. Another illegitimate role occurs 'When scientists suppress, or are asked to suppress, research findings because the results are unpalatable or unwelcome'.[23]

Values can and should play important roles in the sciences, but this need not reflect the loss of objectivity that is sometimes thought to follow. We need only strive towards scientific research that utilizes values in the right ways. We should not use values as reasons for distorting, suppressing or ignoring evidence, conclusions or theories. However, there is nothing problematic about pursuing particular studies because they promise information that society happens to value, nor is it problematic to recognize the social costs associated with the possibility of error in one's work. Finally, as Douglas points out, different sectors of society will not always value the same things, and it is sometimes appropriate to criticize studies because they serve the interests of one group at the expense of society more generally.

The work of Solomon, Hull, Kitcher, Longino, Douglas and others recognizes the sciences as, in part, social activities, that are answerable to broader social concerns. Their ideas give rise to hugely important questions. Are we, as a society, distributing resources and labour sensibly? Are we offering individual scientists the right incentives? Are we excluding some groups and hence important perspectives on important issues? Are we pursuing sciences in ways that reflect what society values, or are our efforts slanted towards benefiting particular groups? All these theorists are aware of, and carefully discuss, particular cases where our research efforts could have been

[23] Douglas (2009, 113).

conducted more responsibly, for example, or where alternative scientific theories were unjustifiably ignored or marginalized. These same thinkers also argue, however, that the sciences are often rational and do often generate substantive, reliable, useful conclusions. Furthermore, their criticisms of particular instances are specific and appeal to more than just the possibility of error. Most importantly, however, since criticism is central to scientific success then we should encourage and seek worthwhile objections to current conclusions, methods and assumptions, but we should also remember that not all objections are equally sensible.

Conclusions

In later chapters, we will adopt a deferential attitude towards particular scientific results, largely on the basis that they are endorsed by the overwhelming number of scientists who have relevant expertise to evaluate the evidence, analyses, inferences and conclusions, and because no good reasons have emerged to seriously doubt the mainstream scientific view, and because the most common objections have been convincingly answered. Nevertheless, we might wonder whether the majority opinion of scientists is one that we are generally obliged to accept.

In response we might suggest that scientists are in possession of the most relevant skills, information, evidence and arguments, and hence it is reasonable to place greater trust in their opinions, but those who seek assurances about given scientific conclusions are unlikely simply to concede this description. Alternatively, we might suggest that those who start evaluating the evidence for themselves will almost certainly reach agreement, eventually, with scientific communities, but this is more an article of faith than an argument. Third, it is apparent that the opinions of experts in other fields, like law, engineering, medicine, finance and so on are afforded special standing, suggesting that a similar attitude towards the sciences is appropriate. However, not everyone places trust in such experts, and there are occasions when it is probably sensible not to.

Of course, that we lack a general justification for accepting the majority opinion of scientific communities shouldn't surprise us. In Chapter 1, we dismissed the problem of demarcation in favour of more direct engagement with the relevant evidence and arguments. If there is no useful feature that distinguishes science from non-science, then there is no feature that achieves special authority for the majority opinions of scientific communities over communities of non-scientists. Chapter 2 helped enforce this conclusion by observing, among other things, that the quality of evidence and reliability of methods of inference are each appropriate objects of analysis. By Chapter 3, we were familiar with the idea that scientific methods can change and that new methods can be introduced. We saw that scientific ideas may be pursued in part for their future promise as well their current success, and that different problems may take on particular significance in ways that make the trajectory of scientific progress hard to anticipate. The complexities, varieties and contingencies of science might again speak against a universal defence. Nevertheless, if we are to adopt certain scientific opinions, then something more positive needs to be said in their defence. We are now in a position to make a few observations.

First, nothing that has been suggested thus far compromises our ambition to identify *poor* reasons for dismissing mainstream scientific results. That my neighbour's grandfather smoked two packs a day, for seventy years, and never suffered a day's ill health, is not a good reason for doubting that cigarettes cause lung cancer. That scientists have been wrong before, that not all scientists agree with the overwhelming majority, that our theories can't explain everything and are confronted with anomalies, that some celebrities doubt certain scientific conclusions, that a significant proportion of society doubts certain conclusions and a great many more fallacious reasons besides are all poor reasons for us to reject those conclusions. Second, of particular scientific issues it is quite within our grasp to start appreciating the basic picture that scientists describe. We don't need to become experts ourselves to become better informed. If we discover that we lack any good reasons for denying particular scientific conclusions, are aware of the basic conclusions described by scientists and keep in mind that the evidence and arguments a layperson comes to appreciate only scratch the surface of what scientific communities accumulate, then we should find ourselves accepting the scientific consensus.

Finally, within this chapter we have reviewed arguments that adopt a more social perspective on questions of scientific authority and that can also help justify more optimistic attitudes towards the conclusions that emerge from epistemic communities that possess particular features. Hull's recognition that scientific communities reward those individuals who correct prior errors and oversights, and similarly social epistemologists' arguments that scientific rationality should be evaluated at the level of communities, might each suggest that scientific communities are better insulated against certain cognitive biases than other groups. Such biases are important not only for evaluating community-based reactions to socially significant scientific issues but also individual reactions. These biases are the subject of Chapter 5.

Discussion questions

1. Philosopher Peter Godfrey-Smith writes that 'speaking roughly, in order to understand knowledge, we need both a theory of human thought, language, and social interaction, and a theory of how these human capacities are connected to the world outside us'.[24] Discuss.
2. Which do you think would make scientific communities more successful at accomplishing their goals – diversity or uniformity of attitudes, approaches, backgrounds, personalities and values?

[24] Godfrey-Smith (2003, 132).

3. Should scientific communities tolerate all conceivable attitudes, values and methods, or should there be limits? What are the limits, what justifies them and who helps establish them?
4. Are you convinced that sciences value novel ideas and theoretical challenges, so long as they are well argued and well supported by evidence? Why or why not?
5. In what ways do you think that past scientific error undermines current scientific conclusions?
6. Would *all* sciences benefit for having more diversity? In what ways? What groups have been excluded from scientific institutions, and what has been the detriment to our understanding?
7. What are the benefits to a community of rewarding individuals who offer compelling arguments and evidence that either correct or extend current understanding?
8. How much should it affect our attitude towards scientific objectivity that scientists pursue only those questions and research programs that are implicitly judged worth pursuing through a combination of individual, community and societal values?
9. When should we be suspicious of scientific consensus? How difficult would it be to evaluate whether the criteria for being suspicious have been satisfied?

Suggested reading

For more on underdetermination see Stanford (2006, 2013). Zammito (2004) traces the history of Quine and Kuhn's ideas through the rise of sociology of scientific knowledge in the 1970s and 1980s, and then on through postmodernism. Psillos (1999) is a great introduction to the scientific realism debate. Kitcher (2003), Solomon (2001), Longino (1990) and Douglas (2009) are all well worth reading.

Points to remember: Part I

1. A satisfying solution to the problem of demarcation remains elusive, but, regardless, we shouldn't be content with dismissing any theory, study or method simply because it doesn't fulfil a particular definition.
2. Empiricist attitudes towards the sciences emphasize both observations and certain forms of reasoning, but the basic empiricist picture could easily lead to confusion about how sciences work.
3. Our experiences are not always reliable indicators of how things are. We should be wary of placing too much emphasis on how things appear.
4. Biases in subjective experience can be overcome through improved experimental design.
5. There might be very good reasons for ignoring certain data, but those reasons may not always be immediately apparent to those who lack relevant training and education.
6. Certainty is not something sciences achieve, but absence of certainty is consistent with having lots of compelling evidence for given conclusions.
7. We continue to learn about how past experiences can better predict future events.
8. Scientific methods are themselves appropriate objects of criticism and refinement, and the introduction of new methods may be entirely justifiable.
9. All theories confront anomalies, so anomalies alone can't be adequate reasons to dismiss a theory. Theories are pursued, in part, because they offer the most promise for purposes of furthering our understanding.
10. Kuhn's suggestion that scientists work within something like a *paradigm* has historical plausibility, but this need not imply strong incommensurability between paradigms.
11. Scientific progress is contingent, shaped in part by what problems scientists choose to engage, what technology becomes available, as well as broader socio-economic and political factors. Nevertheless, such contingencies are quite consistent with scientists achieving substantive results relative to the objectives they set.
12. The arguments from underdetermination and discontinuity within the sciences are each reminders of scientific fallibility. Neither provides compelling reasons to deny *particular*

scientific theories or conclusions that some people may find inconvenient or objectionable.

13. Scientific communities may be a more appropriate object for explaining scientific authority.
14. Sciences are not value free, but the role of values in sciences need not compromise scientific authority.
15. It is appropriate to criticize scientific conclusions, methods and theories, but we should remember that some criticisms may be entirely misguided. Worthwhile criticisms are precise and well-supported.

Part II

Biases, Arguments and Created Controversies

5

Inherent irrationality: cognitive biases and heuristics

When Ola Svenson surveyed American students about their driving skills, 46 per cent estimated that they were among the most skilled 20 per cent, and 93 per cent of those surveyed placed themselves in the top 50 per cent. Doctors who observe several patients that all have the same illness are more likely to diagnosis the same illness in subsequent patients who don't have it, even in the case of noncontagious diseases. Subjects in a study who had been manipulating words like *Florida*, *wrinkle*, *grey* and *bald*, walked away from the testing room more slowly than those subjects who had been manipulating sets of words that don't induce thoughts of old age.[1] These represent just a few of many studies that demonstrate how our decisions, behaviour, memories and beliefs can be influenced in ways that we're wholly unconscious of. Not only are we unaware of these subconscious influences, but in some cases our beliefs are being shaped in demonstrably irrational ways.

In Chapter 2 we discovered that our *perceptions* of the world can be influenced by expectation, among other things. We were reminded that what we perceive doesn't always reflect how things are. In this chapter, we observe that how we reason from available evidence to conclusions and beliefs is also subject to systematic and often predictable error. These errors have enormous significance for our appraisal of those scientific issues where scientific opinion might appear divided. In some instances we're inappropriately influenced by prior commitments in our assessment of new data, or overly influenced by considerations that come more readily to mind, or overconfident in our abilities to understand, predict and adjudicate. One purpose of this foray into cognitive psychology is to forearm ourselves against the possibility that each of us is placing more trust in our individual abilities than is appropriate. A little humility and greater awareness for the ways in which we're vulnerable to error represent key steps towards improving our reasoning abilities.

[1] For driver attitudes, see Svenson (1981). The tendency of doctors is described in Sutherland (2007) and the effects of manipulating age-associated words in Ariely (2010).

Confirmation bias

Confirmation bias is a well-known, widespread and well-studied cognitive bias. It describes our tendency to prefer, or actively seek out, evidence that supports rather than challenges our pre-existing beliefs or desires. Consider Jones, for purposes of illustration, who has a poor opinion of women drivers and who occasionally witnesses instances of reckless driving. If the gender of the driver is unknown, Jones confidently predicts that the driver is female. Subsequent observations sometimes put that prediction to the test. We don't need to keep careful records here, concerning the success of those predictions, because the important feature of our story is Jones's reaction. Insofar as Jones attaches significance to correct predictions, but dismisses failed predictions as outliers that should be ignored, Jones is guilty of confirmation bias. Jones has an opinion about women drivers, and that opinion is preserved in part by ignoring subsequent evidence that challenges the opinion and inflating the significance of evidence that aligns with the opinion.

Clearly, if we allow our existing beliefs to determine the significance of new evidence in this way, then we will tend to retain those original beliefs even if (from a neutral perspective) the new evidence speaks strongly against them. If Jones dismisses all instances of men driving poorly as outliers and aberrations, but regards instances of women driving poorly as critical, telling and decisive, then he (or she) is guilty of confirmation bias. Whether women drivers are better or worse than male drivers is beside the point – it is Jones's attitude towards new evidence that reveals the confirmation bias.

To further illustrate, consider the sequence of numbers '2, 4, 6'. Now consider that the numbers were not selected entirely arbitrarily, but follow a simple and particular rule. Your objective is to discern what the rule is. To do so you are asked to offer further sets of three numbers. Someone who knows the rule will tell you whether your numbers satisfy it or not. Once you're confident that you've identified the rule, you can declare it. If you're wrong, you can continue offering additional sets of numbers, announcing your opinions as and when you become sufficiently confident that you've properly identified the rule. Take a minute to consider what you think the rule might be. Now think of a sequence of numbers you would offer to help discern whether you're right.

If you're like most people, then the rule which first occurs to you is something like 'even numbers that increase in increments of two'. If you're like most people then you'll test your hypothesis by providing a sequence of numbers that satisfies the rule you've selected. Perhaps you offer '8, 10, 12' or '12, 14, 16'. Both these sequences fit the rule that you've been charged with identifying. Finally, if you're like most people, learning that '8, 10, 12' does satisfy the rule increases your confidence that your conjecture was correct. If you're cautious, you might offer another set of numbers that fit what you suppose the rule is, before declaring that you've solved the problem. Importantly, what most people do not do is offer sequences that are inconsistent with their conjecture. Having decided that the rule could be 'even numbers increasing in increments of two', for example, people do not typically think to offer a sequence like '5, 7, 9', which does not fit with their guess, or '4, 5, 6'. In fact, both these sequences also satisfy the rule. The mistake here is one of actively seeking out evidence that is consistent with our hypothesis and failing to seek evidence that could falsify it. Learning that a guess is wrong is within easy reach of subjects, but rarely grasped.[2]

So at least in some circumstances people are more likely to pursue evidence for its potential to positively support their belief, rather than for evidence which, if discovered, would falsify it. We adopt beliefs and opinions, which then influence how we seek or interpret new evidence. Sometimes, as with Jones, we handle evidence differently, according to whether it supports or challenges our pre-existing beliefs. Sometimes, as with the number rule example, we are inclined to actively seek evidence that positively supports our conviction rather than evidence which has the potential to falsify.

Confirmation bias was also nicely illustrated in a study involving subjects who either strongly supported or were strongly opposed to the use of capital punishment. Subjects were presented with fictional articles concerning the efficacy of capital punishment as a deterrent to would-be offenders. The interesting and significant trend among participants was that of holding to higher standards those studies that conflicted with their prior beliefs. Those who supported the use of capital punishment described flaws in the studies that provided evidence for the ineffectiveness of capital punishment, but not those studies that provided evidence to the contrary, and vice versa for those who were initially opposed to capital punishment.

The bias is related to the problem of cherry-picking data. Suppose a large amount of data has been accumulated, in service to the resolution of a particular scientific issue. The vast majority of the data, let's suppose, gestures towards the same basic conclusion. A small minority points in the opposite direction. Other things being equal, we should of course prefer that conclusion which is supported by many studies rather than few. If nine out of ten cats prefer Whiskas, then it would be obtuse to draw attention to an unconventional cat and proclaim that cats don't prefer Whiskas. When individuals' cherry-pick they highlight certain data *because* they support a conclusion they are independently motivated to defend, *despite* those data representing a very small quantity of available data and where the majority of available data actually gestures towards a very different conclusion. Cherry-picking can be intentional. When unintentional to some extent it's a product of confirmation bias.

The tendency to irrationally retain confidence in existing beliefs, despite new and conflicting evidence, has been demonstrated in a wide variety of studies. Let's consider one more. In this example participants were presented with apparent suicide notes and challenged to distinguish genuine suicide notes from those that had been created by the experimenter. Participants made judgements one note at a time and

[2] See Sutherland (2007) for further details and references.

were informed after each guess whether or not they were correct. Unbeknownst to all subjects, however, none of the notes were authentic, and each participant had been assigned at random to one of two groups. Members of the first group were told that they were doing very well at identifying whether a given note was authentic or fraudulent, whereas the second group were informed that they were doing poorly. In the end, everyone was made aware of both the absence of authentic suicide notes from among the selection, and the meaninglessness of being told whether they were right and wrong in their assessments.

What was of greatest interest to the psychologists who designed the study, however, were subjects' attitudes once they'd completed the test. All were asked how confident they were that they could distinguish genuine suicide notes from fakes. Those who had been encouraged during the test that they could make such discriminations were much more confident than those who were informed otherwise, that they could reliably make such distinctions in the future. This was despite the fact that all participants had been informed that their experiences revealed nothing about their ability with this task.[3] Even seemingly trivial beliefs, held for a very brief amount of time, can continue to affect our reasoning such that learning we have absolutely no reason to accept those beliefs doesn't properly restore a sensible attitude towards them. It bears reflecting on how much greater influence certain ideas might have if we've held them a long time, have heard them repeated often, have reasons for wanting them to be true and perceive them as central to broader ideologies.

Once we've reached a conclusion on some issue, our evaluation of subsequent data can become far less reliable. Francis Bacon in the seventeenth century remarked: 'The human understanding when it has once adopted an opinion (either as being the received opinion or as being agreeable to itself) draws all things else to support and agree with it.'[4] The relevance of these biases to certain apparent, scientific controversies should be clear. If someone has formed an opinion about the benefits of homeopathy, or abstinence-only sex education programs, or the putative connection between vaccinations and autism, then these attitudes will affect their reaction to new evidence and arguments. The bias we've described suggests that people will often attach greater weight to evidence that supports their suspicions or beliefs, or look only for examples that provide positive support. If we are to overcome these pernicious tendencies, then we should always be careful to reflect on whether we possess good reasons for our prevailing beliefs, why other (perhaps very well-qualified) individuals seem to reach different conclusions and to consider whether there are better ways of evaluating certain issues beyond those that we're appealing to.

[3] See Sutherland (2007) for further details and references. [4] Bacon (1620, 97).

The availability heuristic

In the year following the terrorist attacks of September 11, 2001, air travel in the United States dipped significantly. Corresponding with the reduction in air travel was an increase in road traffic. Seemingly a lot of people now preferred to drive rather than fly. A national increase in road travel, however, results in more car accidents, and more fatalities and serious injuries suffered in car accidents. One analysis suggested that in the year following the attack, upwards of 1500 more people died in traffic accidents within the United States than would have done if those terrorist attacks hadn't occurred.[5] Part of the explanation for increased road use is probably that some travellers were deterred by the increased security at airports, and hence the increased personal inconvenience. It seems likely, however, that many decisions to avoid flying were influenced by a heightened fear of flying, and that those fears are attributable to the indelible impressions left by the terrible images and reports from the events of September 11.

The pictures surrounding the terrorist attacks were incredibly distressing, overwhelming for many, vivid and memorable for everyone. Images and thoughts that make a profound impression upon us are far easier to recall. Decision making and belief formation that's influenced by what's most *available* to us is a product of the availability heuristic. Poor risk assessment is one consequence of this cognitive bias. Flu shots are extremely safe. Nevertheless, if you know someone who became ill soon after having had a shot, this is likely to influence your decision whether or not to have the shot yourself. The story of one individual *shouldn't* affect our decisions, but empirical studies demonstrate that it does.

When the availability heuristic was first described, it was understood as affecting our ability to reliably evaluate the frequency of certain types of event. If it is easier to think of examples of accidental death versus deaths from diabetes, then we tend to assume that the former are more common. The heuristic can lead us astray, since the reasons that certain types of event come to mind more readily than others will not always reflect their relative frequency. Some stories receive far more media coverage, or are of great personal importance, or are relayed in a more memorable and vivid way. A well-told, heartfelt story about a mother's struggles with her autistic child will become very available to us, and hence we may misjudge the actual frequency of autism because cases that we've heard are easy to remember. Importantly, however, such stories will also be more available to us than statistics describing the number of lives that have been saved as a result of a vaccination program, or official statements from medical societies and professionals which describe the excellent safety record of vaccinations. The power of one story, coupled with even a highly dubious hint that autism is causally related to vaccinations, or anything else for that matter, might suffice to force far more significant information out of mind and induce decisions that

[5] Gigerenzer (2006).

are irrational and dangerous. In this example, availability doesn't affect judgements of relative *frequency*, but crowds out what are actually more salient facts. The concept of availability can thus be extended to account for additional ways in which we err in our reasoning.

The ease with which certain ideas come to mind influences our judgements concerning the frequency of those events and, more generally, their significance. In one study, subjects were given a list of six words and told that these words described a person.[6] They were asked to provide a more detailed description of what this person was like, given the six words they'd already been given. All subjects received the same set of words, but the order of the words was reversed for half the subjects. One group was told that the person was *intelligent, industrious, impulsive, critical, stubborn* and *envious*; the remaining subjects were told the person was *envious, stubborn, critical, impulsive, industrious* and *intelligent*.

Those subjects who had received the list with the more positive adjectives at the beginning of the list offered considerably rosier descriptions. One plausible explanation offered for these results is that whichever words appeared earliest in the list, being so available to the subject, shaped their attitude towards those words that appeared later in the list. If something as trivial as the order in which ideas are presented can shape the way we process information, then we should clearly be on our guard against the effects of highly emotive, personal stories that may be widely discussed in the media.

What's also relevant to the availability heuristic is the fact that some ideas are transmitted more successfully than others. We remember certain ideas more easily and feel more inclined to share certain stories and experiences over others. Our collective habits in these respects help explain the spread of urban legends, public health scares and amusing, bizarre, perhaps disgusting pieces of trivia. The success of ideas, in terms of their capacity to spread, often has little to do with their truth. Mark Twain seems to have noticed this when he quipped that 'a lie can travel half way around the world while the truth is putting on its shoes'. The *reasons* that certain ideas propagate more successfully are important, and we will return to these shortly. However, the simple fact that some ideas receive more attention is also significant.

Mere repetition renders beliefs more available. A given claim is expressed, despite lacking any real merit, support or evidence; the more it gets repeated, the more people are exposed to it; the more that people are exposed to it, the more available the idea becomes. The availability heuristic leads us to expect that as ideas are repeated more, so they are likely to be regarded more favourably. This prediction has been studied, and confirmed. It has become known as the bandwagon effect.

In the 1950s, Solomon Asch conducted some famous experiments in social psychology. In one such study a group of subjects was presented with pairs of cards. One card from each pair had a single line drawn on it; the second had three lines of

[6] Described in Kahneman (2013, 82–85).

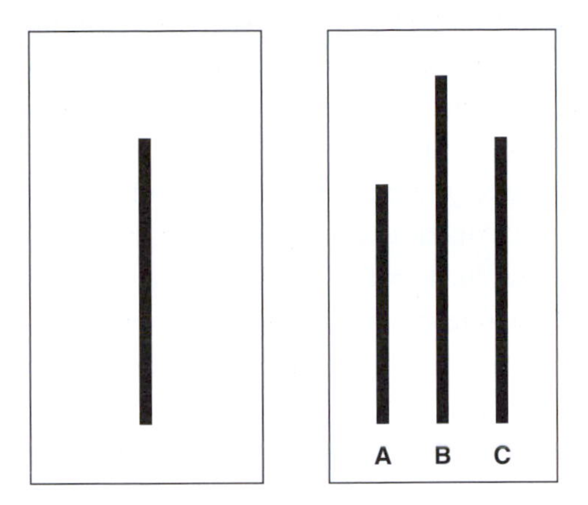

Figure 5.1 Example of Asch's cards. Which line on the right card is equal in length to the line on the left card?

differing lengths (see Figure 5.1). Participants were asked to judge which of the three lines was identical in length to the line on the first card. Control experiments had earlier suggested that identifying the correct answer was not difficult. The pairs of cards were now presented to groups of between six and eight people, but all except one had received prior instructions to give some false answers. The real subject of the experiment was the poor patsy whose observational evidence was unambiguous and compelling, but who heard several members of the cohort agree that the answer was something else. Would subjects really deny the evidence of their own senses and knowingly offer a false report for the sake of conforming? Asch found that 75 per cent of subjects gave at least one incorrect answer and that almost a third of all answers given by the subject were incorrect on those occasions when the actors gave incorrect answers. Asch's experiments have been repeated, with similar results. Participants in these studies, however, often deny having been influenced by the group. A more recent study concluded that people tend to 'infer that a familiar opinion is a prevalent one, even when its familiarity derives solely from the repeated expression of one group member', and that the effect 'holds even when perceivers are consciously aware that the opinions come from one speaker'.[7]

The bandwagon effect presents a clear opportunity for those who would distort scientific conclusions or promote unsupported ideas. If you can get your idea repeated enough times, then this alone increases the credence that people will attribute to it. Ensuring that more media outlets are picking up the story, and that your websites are attracting more traffic, will increase the volume of reports and discussions, and this alone will help promote support for your cause.

[7] Weaver et al. (2007).

The fact that repetition increases availability is itself significant. Reflecting a little on the reasons that some ideas spread more efficiently is also worthwhile. Chad and Dan Heath in *Made to Stick* consider the question of why some ideas are *stickier* than others.[8] They argue that several features of ideas increase their stickiness. Simple, profound ideas are more likely to be remembered and shared, as are ideas that seem plausible according to the standards of the audience. Attaching emotional content to an idea increases stickiness, as does the use of illustrative stories rather than statistics and data. At least some of these reasons for increased stickiness help illustrate the challenge of defending scientific conclusions against stubborn, well-motivated opponents.

Scientific theories are complicated sets of ideas. Some aspects of a theory are supported by some data, but other aspects might require a different source for their confirmation. Certain aspects of a theory are more central to a research program than others. The world is complicated and, as a result, so are our scientific theories, our methods for gathering and analysing data and evaluating conclusions and hypotheses. Scientists devote enormous amounts of time and energy towards understanding the methods, techniques, arguments and results of their particular discipline. When scientists seek to engage their detractors the complexities may surface. But now we should pause to consider which is most memorable, striking, and vivid: the technical details of a scientific explanation, populated by jargon that we might not fully grasp, independent claims for which the justification might be quite opaque, cautionary remarks about the limits of understanding in certain respects and so on, or the short pithy objection whose content we grasp immediately. If certain objections to mainstream science are more *available*, because they possess the features of sticky ideas and not because they're salient, then we'll remember the objections better than the answers, and hence our judgements about the state of the science will be less reliable.

Overconfidence

A bat and ball cost $1.10. The bat costs one dollar more than the ball. How much does the ball cost?

It takes five machines five minutes to make five widgets. How long will it take 100 machines to take 100 widgets?

In a lake, there is a patch of lily pads. Every day, the patch doubles in size. If it takes 48 days for the patch to cover the entire lake, how long would it take for the patch to cover half of the lake?

These puzzles involve nothing more than basic arithmetic, yet when thousands of university students were presented with the questions over 80 per cent of students got

[8] Heath and Heath (2007).

at least one question wrong and one third got all three wrong. Even at highly selective universities, like Harvard, MIT and Princeton, 70 per cent of students who took the test gave at least one incorrect answer.[9] Even those who get the answers correct notice something about each question: there is an intuitively appealing answer to the puzzle and then there is the correct answer. An answer pops into mind (10c for the first question); it's straightforward to check our answer (if the ball costs 10c, then the bat costs $1.10, given the information provided in the puzzle, but then the total cost would be $1.20, so the ball can't cost 10c after all). Why do so few people check their answer? Daniel Kahneman, a Nobel Prize-winning psychologist, describes the problem here as involving *overconfidence*. We subconsciously suppose that if an answer *feels* right, then it probably is right. We have too much confidence in our instincts or intuitions.

Overconfidence has also been neatly illustrated by studies where participants complete a task and then describe their level of confidence that they were successful. When asked to spell a word, those who were 100 per cent confident that they spelled it correctly were wrong 20 per cent of the time. We're also often overconfident in our abilities, relative to our peers: at the University of Nebraska, 68 per cent of their faculty judged themselves within the top 25 per cent at that institution for teaching ability. Kahneman relates overconfidence to several additional cognitive biases. For example, various studies illustrate that we are overly confident in our abilities to understand the past. Once blessed with the benefit of hindsight, we regard events as more probable, given past available evidence, than we do when we are ignorant of whether the event occurred. The hindsight bias has costly social implications, but for our purposes a more important manifestation of overconfidence concerns our penchant for *coherence*.[10]

Our understanding of events and issues is fuelled, necessarily of course, by the information that we have available to us, but what's available is incomplete and often of poor quality. Nevertheless, we attempt to accommodate the information that's available within a coherent, simple story. Furthermore, various studies suggest that we're acutely inattentive both to the information gaps and the quality of the evidence that's available. Even when subjects are explicitly advised that there is further information which is not being disclosed, or that the information provided is unreliable, there's still a tendency to hang significant weight on what is available. Statistical information is also sacrificed in our pursuit of a good story. The complete absence of evidence for certain beliefs is sometimes no obstacle if those beliefs *seem* plausible to us.

To illustrate, Kahneman draws attention to an example from the statisticians Howard Wainer and Harris Zwerling regarding the rate of kidney cancer across the United States.[11] They provide a map that represents the kidney cancer rate for different counties and observe that those counties with the very lowest rates are

[9] Described in Kahneman (2013, 44–45). [10] For more on the hindsight bias, see Kahneman (2013, 201–204).
[11] Described in Kahneman (2013, 109–110).

very rural counties. They then observe how: ‘It is both easy and tempting to infer that this outcome is directly due to the clean living of the rural lifestyle – no air pollution, no water pollution, access to fresh food without additives and so on.’ They also note, however, that those counties with the very *highest* rates of kidney cancer are also rural counties. Wainer and Zwerling now observe: ‘It would be easy to infer that this outcome might be directly due to the poverty of the rural lifestyle – no access to good medical care, a high-fat diet, and too much alcohol and tobacco.’ But living in rural areas can’t both increase and decrease the chances of developing kidney cancer, so what are we to make of the data? In fact, it’s all about sample size.

Across the population of the United States the rate of kidney cancer is about five in every 10,000. In highly populated counties the rate will always fall close to the national level. In counties with low populations, however, the cancer rate can deviate from the national average much more significantly and very easily, but there is no causal story to be found here concerning the risks and benefits of rural living. The explanation is mathematical or statistical. Consider a county with less than one thousand inhabitants. If just one person within the county develops kidney cancer, then it will immediately have a kidney cancer rate that far exceeds the national average. However, if one person fewer develops lung cancer (that is, no-one in this particular county), then the rate is zero and hence the lowest in the country. If we’re concerned with the frequency of some property within a certain population, then the smaller the sample size the more likely it is that we’ll observe frequencies that depart significantly from the frequency of the entire population. Ignoring this statistical fact, and trusting instead the intuitively appealing stories that we find it easy to construct, is one important way that we can bungle in our thinking. Our subjective confidence in coherent stories leads us to ignore important statistical information. We might possess overwhelming evidence that only a very small proportion of a given population will achieve some particular goal, but of those who are successful we believe we understand the *reasons* for their success. We attribute far too little to chance and luck.

As well as ignoring sample size we are insufficiently attentive to base rates. Suppose a speed-gun is positioned on a stretch of road where one in every thousand drivers is known to exceed the speed limit by at least 15 mph. The speed-gun, let’s suppose, reliably takes a photo of all such vehicles, but also photographs about 1 per cent of vehicles that are not exceeding the limit to such an extent. If we learn that a particular car has been photographed, how likely do we suppose it is that the car was driving at least 15 mph over the speed limit? Any tendency on our part to suppose that it is very likely is an illustration of our susceptibility to the *base rate fallacy*. Perhaps our problem is that the provided information suggests that the speed-gun is very reliable – it catches everyone who’s guilty and correctly ignores 99 per cent of those who aren’t. What we miss, however, by attending only to these two pieces of information, is the all-important base rate – the rate at which speeding to the specified degree occurs within our population. Reflecting on the base rate soon reveals why our initial reaction is spurious. If 999 out of 1000 drivers on this stretch of road are driving within 15 mph of the limit, but 1 per cent is captured by the camera nonetheless, then on average ten out of every thousand drivers will be needlessly photographed and only one driver out of every thousand will be photographed who is actually speeding. For every eleven cars that are photographed, therefore, only one is driving at least 15 mph over the limit.

Overconfidence likely has a number of consequences for evaluating the state of scientific debates. Perhaps we overestimate our personal understanding of the scientific theories under discussion, or our ability to evaluate conflicting arguments and evidence, or our ability to ensure that our beliefs aren't being swayed by emotion or wishful thinking. We are likely greatly influenced by what intuitively seems plausible, rather than by what the evidence suggests. Although we're aware that relevant experts are in possession of far, far more relevant information than we have even heard about, much less can recall, we will underestimate the significance of that which we're missing. And, in many circumstances, we allow reliable statistical information to be irrationally trumped by appealing narratives.

Memory

There are many liabilities that psychologists have uncovered which concern aspects of our memory. New information can influence the way we recall past events, for example. In one classic experiment, participants were shown images that included a green car driving past the scene of an accident. They were then asked questions about the images. One of those questions falsely implied that the car was blue. When subsequently asked about that vehicle, many recalled it being either blue or bluish green.[12] Other studies suggest that the likelihood of us accurately recollecting something can be altered by situation, mood and desires. The number of memory-related effects, biases and liabilities is legion. Here I'll draw attention to a couple that I hope readers will remember later in the book.

First, the selective nature of memory leads us to suppose that illusory correlations are genuine. Those occasions when we were leaving the house in an unusual hurry and were unable to locate the car keys might stand out in our memory, leading us to suppose that the keys always go missing on precisely those occasions when it is most inconvenient. We run out of things, break things or need to replace things – all at the worst possible times. Or, at least, it often feels like that. Psychological studies demonstrate that often we confuse genuine with illusory correlations, thus tend to look for causal relations where there are none. A second memory-related bias is the tendency to believe claims that one remembers having heard before, even on those occasions when claims were initially introduced for the sole purpose of being

[12] Loftus (1977).

discredited. In a study at the University of Michigan, for example, subjects were introduced to various claims and informed, either once or repeatedly, that the claims were true or false. Three days later, older subjects were more likely to misremember false claims as true if the claims had been mentioned repeatedly. Repeating information for the sole purpose of discrediting it can have the opposite effect.[13]

Decoy effects

In *Predictably Irrational*, Dan Ariely describes and illustrates a kind of irrationality that's associated with decision making. Sometimes, we're confronted with easy choices, and sometimes we're confronted with hard ones. Ariely argues that, across a range of circumstances, introducing an easy decision alongside a hard decision has a predictable but irrational effect on how we behave. Here's one of Ariely's illustrations:

> Suppose you are planning a honeymoon in Europe. You've already decided to go to one of the major romantic cities and have narrowed your choices to Rome and Paris, your two favorites. The travel agent presents you with the vacation packages for each city, which includes airfare, hotel accommodations, sightseeing tours, and a free breakfast every morning. Which would you select? (Ariely, 2010, 10)

The decision might be a hard one, but now, Ariely continues, suppose a third option is made available. It's exactly like the Rome package except that it doesn't include the free breakfast. Clearly, if you're going to Rome you may as well have the free breakfast. The choice between Rome with or without a free breakfast is an easy one. What the decoy effect describes is our tendency to let the easy decision influence our attitude towards the hard one. There is no good reason why introducing the choice of Rome without free breakfast should influence us to choose Rome with free breakfast over Paris with free breakfast, but various studies conducted by Ariely and others strongly suggests that it does.

One such experiment concerned an advertisement for subscriptions to *The Economist*. The information that was presented in the advertisement which caught Ariely's attention is summarized in Figure 5.2.

Subscription options for *The Economist*:	
One-year subscription to Economist.com (which includes online access to all articles from *The Economist* since 1997)	$59
One-year subscription to the print edition of *The Economist*	$125
One-year subscription to the print edition of *The Economist* and online access to all articles from *The Economist* since 1997	$125

Figure 5.2 Information taken from an advertisement printed in *The Economist.*

[13] Schwarz et al. (2005).

What's immediately apparent is that the combined *Print and online* option is much better value that the *Print-only* subscription. The pricing seems strange, but afforded Ariely an opportunity to test the decoy effect. The choice between the print-only subscription and the combined subscription is an easy one. If you're going to spend $125, you may as well get more rather than less. The choice between the combined package and the online-only subscription, however, is a harder one. Perhaps having the magazine delivered to your door increases the chances you'll read it, which might motivate some to spend more to have the print copy. Some people prefer reading from paper, rather than online. The difference in price, however, is not insignificant. The decoy effect thus predicts that introducing the combined subscription package at the same price as the print-only subscription, and thereby introducing an easy choice, will influence people's decision.

Ariely tested the prediction on 100 MIT students. When presented with all three options, 16 students preferred the online only-subscription, while the remaining 84 chose the combined package. Quite understandably, no-one wanted print-only. To test the strength of the decoy effect Ariely then offered 100 students just the hard choice. That's to say, the print-only option was completely removed so that students could choose either the online-only subscription or the combined package. Of course, since the only option that had been removed was the one that no-one in the first study had selected (when subjects had been presented with all three options), then there was no rational reason for the proportions of voters to change significantly. If the combined package was more desirable than the online subscription, as the first poll strongly suggested, then that same preference should have been apparent in the second poll. In fact, 68 of the 100 polled students this time preferred the online-only subscription, over four times as many.

A related phenomenon is known as the compromise effect. This is often described in terms of retail consumer behaviour. When consumers are presented with two alternatives, their choices may split fairly evenly, but if a third option is introduced which makes one of the original choices into a kind of compromise, then that compromise option becomes more desirable. For example, suppose consumers can choose between an inexpensive television with relatively few functions, and a more expensive model with more features. The compromise effect describes how the proportion of expensive models sold will increase if a third model is introduced into the market that is yet more expensive and has even more features. What has now become the compromise position, relative to all three models available, is the most appealing to consumers.

How do the decoy and compromise effects relate to scientific controversies? Here's what I have in mind. When science presents us with conclusions that are disturbing, inconvenient or otherwise disappointing, we may feel pulled in two directions. Deciding whether to accept or resist the science can be a hard decision. There is motivation to deny the science, but many of us will be concerned that such denial makes us irrational, and hence that we will make bad decisions and look ignorant and foolish. It is into this predicament that strange hybrid beliefs might look attractive. For example, biologists accept that species evolve and change over time, but some sceptics maintain that species are fixed and immutable. Deciding which attitude seems most plausible might be a hard decision. Now suppose you're informed of a distinction between macroevolution and microevolution, and informed that the distinction plays a role within contemporary biology. Microevolution describes small-scale changes *within* species that can be observed over relatively short periods of time, either in the laboratory or the field. Macroevolution refers to changes that take much longer to unfold, and is often defined as evolutionary change at or above the level of species.

The consensus among biologists is that while distinguishing microevolution from macroevolution is useful for addressing certain questions, there is no doubt that the evidence for microevolution and macroevolution is extremely compelling and of generally the same character. A lot of very compelling evidence speaks in favour of macroevolution just as much as microevolution. Available evidence does nothing to warrant accepting only the idea of microevolution. However, if a religiously motivated Creation scientist suggests that the evidence for microevolution is more compelling and that we should accept only the idea that species change just a little bit, then this could function as a decoy effect. The choice between the immutability and evolution of species might be a hard one, but the choice between supposing species change a little bit and supposing they change a great deal is easy. Belief in microevolution, but not macroevolution, becomes appealing. The appeal might have very little to do with an individual's understanding of the evidence for microevolution, macroevolution or immutability – it's a product of the decoy effect. More generally, the decoy and compromise effects might help explain why we feel some compulsion to suppose that there's probably some merit to both sides of a given debate. Deciding whether to endorse or repudiate a given claim can be a hard decision – defaulting to a middle position can improve cognitive ease, whether or not the middle position is a reasonable one.

The decoy and compromise effects are examples of a more general cognitive bias whereby we seemingly replace one question with another, without noticing we've made the substitution. Daniel Kahneman discusses what has become known as the *affect heuristic*, as a further example of such subconscious substitutions. Phrases and words like *nuclear power*, *genetically modified foods*, *vaccinations*, *abortion* and so on provoke an emotional response in many people. Studies suggest that our emotional responses influence our assessment of the risks and benefits associated with such procedures and technologies: when people are presented with information extolling the benefits of certain technologies, they come to regard them as less risky, even though they've been presented with no evidence about the risks; likewise, evidence that technologies are relatively risk-free produces more positive evaluations of their benefits.[14] The role of

[14] See Kahneman (2013, 367) for more details.

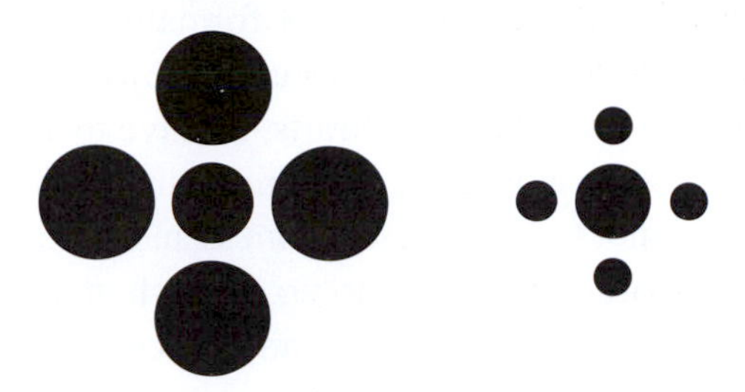

Figure 5.3 Another optical illusion.

emotions within decision making is also well illustrated by a phenomenon known as *emotional framing*. When Harvard physicians were presented with statistics concerning the effects of radiation versus surgery for purposes of treating lung cancer, their responses differed markedly depending on how the information was presented. For some, surgery was described as having 90 per cent *survival rate* in the first month; others were informed that surgery involved a 10 per cent *mortality rate* within the first month. The information is the same, but physicians responded differently.[15]

Bias blind spot

Our glimpse into the fascinating world of cognitive psychology and behavioural economics is almost over, but one further feature of cognitive biases and heuristics deserves particular emphasis. It is a feature that has been illustrated by analogy with optical illusions. Consider the circles at the centre of the two designs in Figure 5.3. The central circle on the design to the right appears larger than the central circle on the left hand design. It isn't. Hide the four surrounding circles from each design, and the illusion disappears. What's striking about this and many other optical illusions, however, is that when we unveil those outer circles the illusion reappears. We can convince ourselves that our perception of some image is misleading, but that doesn't affect the actual perception. Even once we know that the circles are the same size, we perceive them as different. Something similar is frustratingly true of at least some cognitive biases. Even once we're made aware that certain kinds of judgement are irrational, we remain susceptible to those same irrational tendencies. We learn that subconscious influences shape people's choices, and that these sometimes result in less than optimal outcomes. We acknowledge these in the abstract, but in our own particular circumstances we fall into the trap of thinking again that our reasons are fully known to us, that our conclusions are reasonable and that *in this instance*, at least, no subconscious influences are acting. We do well to remember that subconscious influences are always present *even if we're not aware of them!*

[15] Described in Kahneman (2013, 367).

This feature of cognitive biases has also been formally investigated. In one study, overseen by Princeton psychologist Emily Pronin, subjects were asked to evaluate themselves with respect to six personality traits, relative to a well-defined cohort.[16] Of these six traits three were admirable (dependability, objectivity, consideration for others) and three were not (snobbery, selfishness and deceptiveness). The results of the surveys nicely illustrated what's known as the better-than-average effect, a type of overconfidence bias that we've already discussed. The effect describes people's tendency to evaluate themselves more positively than they evaluate others. When asked to compare themselves with some given population, subjects rank themselves as more virtuous, talented, compassionate and so on than the average member of that population. In Pronin's study, in keeping with the better-than-average effect, the majority of participants considered themselves more dependable, objective and considerate than average, as well as less snobbish, less selfish and less deceptive. The main purpose of the study, however, was to learn how the subjects would react once *informed* of the better-than-average effect. To this end, subjects were presented with a short description of the effect and asked to consider how their own self-evaluations, which they'd just presented, would compare with more objective measures of these six personality traits. In particular, they were asked whether:

> The objective measures would rate me *lower on positive characteristics* and *higher on negative characteristics* than I rated myself
> The objective measures would rate me *neither more positively nor more negatively* than I rated myself
> The objective measures would rate me *higher on positive characteristics* and *lower on negative characteristics* than I rated myself.[17]

The first alternative reflects individuals' recognition that they are themselves probably susceptible to the better-than-average effect, hence likely to evaluate themselves overly favourably, relative to the rest of the population. The second alternative represents individuals' unwillingness to admit that they were vulnerable to the effect, at least on this occasion. The final option suggests that not only do the participants deny they are biased in this particular respect, they actually think that, on reflection, they rated themselves more modestly than would an objective set of measures. So how did participants answer this follow-up question? Around three-quarters denied that their evaluations had been influenced by this effect, despite being told that 70–80 per cent of subjects rate themselves above average. In virtue of this and additional studies, Pronin and her collaborators suggest that, 'knowledge of particular biases in human judgement and inference, and the ability

[16] Pronin et al. (2002). [17] Ibid, 375, original emphases.

to recognize the impact of those biases on others, neither prevents one from succumbing nor makes one aware of having done so'.[18]

The phenomenon is known as the bias blind spot. People learn of cognitive biases but, without any good reason for doing so, regard themselves as less susceptible, or even completely immune. The bias blind spot gives rise to an obvious question. If *awareness* of particular cognitive biases isn't always sufficient to improve our reasoning, then how can we avoid succumbing to our irrational tendencies? Some psychologists appear pessimistic about the prospects of overcoming these habits. Others have explored strategies for improving our reasoning that appear somewhat successful, at least relative to particular biases, although it seems likely that distinct biases will require distinct strategies for reducing their effects. Certainly, the best we can reasonably hope to achieve is some mitigation.

Beyond these observations, I still find myself clinging to the, perhaps naïve, hope that surely better instruction can help. Hard though it may be, surely we are each capable of better recognizing, appreciating and internalizing the fact that, if studies suggest that most people are susceptible to a given, irrational tendency, then this itself counts as a very good reason for supposing that we, personally, are susceptible to those same biases. These habits can't be identified through introspection. The best evidence concerning our own susceptibility is the fact that most people are susceptible. Of course, the bias blind spot suggests that we are not good at appreciating this lesson. Perhaps, however, by understanding both the bias blind spot, as well as the core idea behind other biases, we can help enforce the lessons we should be drawing from the cognitive psychology literature: although hard to appreciate and something we are reluctant to concede, *if most people are prone to certain cognitive biases, then it is likely that we are also prone.* The best available evidence encourages us to start with the assumption that we are typical in terms of cognitive abilities, and to revise that judgement only slightly and only if we have compelling reasons to do so. Recognizing this should motivate much closer attention to questions like these:

- Do I have reasons for *wanting* the evidence to point in a particular direction?
- Are there more reliable ways of generating data than those I'm relying upon?
- Are emotional considerations crowding out more relevant information?
- Is my *memory* of certain events or data clouded by what I hope is the case, or what I now judge more plausible?
- Am I placing too much weight on too little data, perhaps of dubious quality?
- Am I placing a level of confidence in my own abilities that is not sensible?

It's not practical to review these questions every time we're faced with decisions. If we care about reaching sensible assessments on those issues that matter to us, however, then cogitating carefully on questions like these becomes essential.

[18] Ibid, 378.

Conclusions

In the last chapter we introduced some of Miriam Solomon's important work on social epistemology, work which makes explicit that individual scientists are prone to various cognitive biases and heuristics, a claim for which Solomon provides evidence. For example, she describes several early-twentieth-century biologists and geologists who were disproportionately influenced by evidence which they experienced directly, rather than merely read about, and thus which was most available to them.[19] Solomon suggests that confirmation bias plays a role for individual scientists' acceptance of scientific theories, as do peer pressure, pride, and deference to authority. More recent work has revealed further respects in which the irrational tendencies of human reasoning may compromise, in specific respects, the objectivity of scientific reasoning. Evidence that such biases influence scientific practice serves as motivation to improve those practices.[20] Solomon's primary interest, however, is to regard the presence of biases as a motivation to evaluate rationality and related concepts at the level of scientific communities. Now that we are better acquainted with such biases it is worth returning to the idea, and Emily Pronin's work is again pertinent.

One of Pronin's studies was aimed explicitly at the possibility that we are better at identifying cognitive biases in others than we are at identifying them in ourselves.[21] This is what the bias blind spot would lead us to expect, and Pronin's research confirmed the prediction. Notably, therefore, a system for generating results and conclusions that requires work to be reviewed by qualified experts is likely to be better at unearthing individual biases. When scientists submit their work for peer review, they are improving the chances that biases will be observed and that corrections can be made. Evidence that we are better at detecting the effects of cognitive biases in the reasoning of others than we are in our own thus fits nicely with the idea that scientific communities are more reliable sources of knowledge than any of their individual members.

Similarly, in light of psychological evidence that humans are vulnerable to confirmation bias and the bandwagon effect, we have reason to prefer systems of investigation that motivate rigorous and compelling defences of novel ideas and corrections to pre-existing commitments. As we noted in Chapter 4, it appears that scientific communities possess this feature. Insofar as scientific communities are better than their rivals at motivating the pursuit of these goals, we have reason to prefer the conclusions they generate. The fallibility of memory is improved simply by keeping careful written records, as are concerns driven by availability. The extent to which the communal dimension of the sciences adequately overcomes the effects of cognitive bias among working scientists is a topic that deserves further attention. No doubt, as we saw in Chapter 4, there will be instances where we conclude that science

[19] Solomon (2001, 61–62). [20] See Elliott and Resnik (2015) for a short introduction to this literature.
[21] Pronin et al. (2004).

could be done better, but to repeat what by now should be a familiar refrain, the worthwhile criticisms of scientific practice draw attention to specific and particular shortcomings rather than the mere *possibility* that widespread prejudices *could* be epidemic across all sciences.

Our reasons for advancing particular conclusions and opinions are often less apparent to us than we're inclined to think. Sometimes, this results in behaviour that is harmless and perhaps even amusing. Another Dan Ariely experiment revealed that diners prefer to choose something from the menu that hasn't already been selected by one of their dining companions, but tend to be less happy with what they end up with. Some studies reveal biases that are just baffling. There's apparently a tendency to bet more money on the role of dice that haven't already been rolled, over dice that have been rolled but about which we're ignorant of the outcome. Do we think we have some control over the future? It's hard to find any other reason for this particular proclivity. Sometimes cognitive liabilities induce outcomes that are far more disconcerting. There are studies which suggest that job applicants with *white-sounding names* are more likely to receive invitations to interview than applicants with *black-sounding names*. Essays are graded more favourably by examiners, if they believe the essays are written by males. Implicit biases are deeply troubling, important to be aware of and hugely important to try to overcome.

As with all sciences, the conclusions of cognitive psychologists and behavioural economists are vulnerable to modification. Competing explanations might be advanced for some experiments. Some biases are more common among some groups, whether distinguished by gender, age, socioeconomic conditions or cultural background. Our understanding of biases will improve, as will our appreciation for how they are best overcome. Despite the fallibility of the results, however, it would clearly be extraordinarily naïve to suppose that our reasons for preferring certain conclusions are completely transparent to us. When we pause to consider *why* some people believe what they do, we should recognize a crucial ambiguity in the question. The reasons that people *offer* in defence of their opinions won't always align with their subconscious reasons. In subsequent chapters, we will return to this distinction.

Cognitive biases and heuristics don't just render our reasons for holding certain ideas less transparent to us, they also result in poor evaluations of available evidence and hence irrational beliefs. Where possible, we should seek more objective data and express more caution in the conclusions we are willing to draw. In some cases, the superiority of scientific efforts is blatant. To establish whether average global temperatures are increasing, I could think back over the summers of my childhood for evidence that there were fewer hot days, or more recent memories for evidence that winters were milder and summers more extreme. It hardly needs saying, however, that our personal experience of the climate is vastly inferior to the colossal data sets that are available to climate scientists. Rather than trust what's *available*, we should seek out more reliable information. Similarly, the layperson's grasp of esoteric details of medicine, anatomy, ocean-atmosphere coupling relations, radiometric dating and so on is meagre. Confronted with rumours and particular claims surrounding these topics, we could speculate about their plausibility, drawing on the few details we can remember, constructing a story that *feels* plausible and trust that story is at least reasonably accurate. Alternatively, we could admit that those thousands of experts who have devoted enormous quantities of time and energy to better understanding these issues, who are familiar with far more sophisticated methods for gathering data, analysing the results and drawing justified conclusions, who have studied these topics from multiple perspectives, and thought about them every day for years on end are collectively far better positioned to reach sensible conclusions.

Some people like to be contrary and will argue for conclusions simply because they're unpopular or outlandish. Some are drawn to grand conspiracy theories or find something gratifying about the idea that they're part of a select crowd who have identified the truth while everyone else remains confused. Denial is a natural, healthy psychological response to diagnoses of serious illness. Hope and desire help explain why many people place more confidence in certain alternative therapies than is sensible. Clearly we could spend a lot of time thinking about psychological explanations for local or wide-scale scepticism or belief formation, but such topics will have to wait for another time.

Discussion questions

1. To what extent do you think people place more confidence in their own ability to evaluate certain scientific issues over the combined wisdom of thousands of experts? Are people aware they are doing this?
2. It has been suggested that people's online behaviour can create *echo chambers*. What are these, why might they be troubling and how are they related to cognitive biases?
3. What strategies might we employ to overcome cognitive biases? Are there any reasons to be optimistic that these would work? How can we evaluate their efficacy?
4. To what extent can we overcome our individual biases by placing more trust in the conclusions of communities of experts?
5. Might rewarding members of a community who discover mistakes in previous theory and research help overcome the bandwagon effect?
6. Might placing a premium on *testing* scientific hypotheses help overcome confirmation bias?
7. Are there other specific ways in which a community of thinkers might be less vulnerable to cognitive bias?
8. Can you think of ways in which communities of thinkers are more prone to certain biases? Are there ways to overcome them?

Suggested reading

Ariely (2010) and Sutherland (2007) are good, accessible introductions to the literature on cognitive biases and heuristics. Daniel Kahneman has been a leader in this field since its inception. His contribution for a wide audience is fantastic (Kahneman 2013).

6

Thinking more clearly: arguments, reasoning and informal fallacies

Central to any scientific endeavour are efforts to advance evidence and reasons in support of particular conclusions.[1] Some conclusions are broad, ranging over all space and time; others are far narrower and may concern the attributes of a particular biological species, chemical element or geological feature. Some conclusions may depart from previous work, requiring some degree of revision to existing attitudes; other conclusions extend our understanding into new terrain. Attempts to understand scientific methods can be understood in terms of a desire to better articulate the *kinds* of arguments that scientists develop, the kinds of evidence that they appeal to and the reasons those methods and evidences are valued.

Very often the ability to comprehend and evaluate scientific arguments requires significant training in the relevant discipline. Gathering data and evidence in ways that are reliable is not always straightforward. Evaluating someone's analysis of the evidence and the plausibility of the conclusions she draws may require familiarity with sophisticated mathematics and statistics, computer modelling or the intricacies of technical apparatus that laypersons typically don't possess. Nevertheless, if we understand science very broadly as the collection of our best methods for investigating the world and reaching reliable beliefs about it, then we would all think more *scientifically* if we spent time developing basic critical thinking skills.

The need for more responsible argument evaluation can also be motivated by reflecting on the cognitive biases that we considered in Chapter 5. Given our natural tendencies to prefer evidence that fits with prior beliefs, for example, it becomes all the more important that we cultivate a habit of pausing to reflect on the reasons we have for holding certain opinions. Do we have good reasons for doubting anthropogenic climate change, or have we merely heard sceptical attitudes repeated sufficiently often that they have attained an inflated and false level of plausibility? An important step towards improving our critical thinking involves asking ourselves what reasons are being advanced in support of some claim, whether those reasons are

[1] This isn't a good way of *distinguishing* science from other forms of investigation, since it's too easily satisfied, thus too inclusive.

themselves plausible and whether there are better ways of accounting for the conclusions being presented. Many arguments are advanced in opposition to mainstream sciences which are so patently unconvincing that basic critical thinking should be sufficient to expose their flaws. In this chapter, we'll review a variety of ideas that can help us all improve the way we assess arguments and debates.

Arguments and statements

It is useful to distinguish between two approaches to the subject of reasoning and argument evaluation. Informal logic confronts and engages arguments as they appear in their natural habitats, in the media, in books, on blogs and in conversation. We inspect the kinds of arguments that are being made, attend to their strengths and weaknesses and in some cases identify and diagnose certain errors that are being committed. When a type of error is made sufficiently frequently, we give it a name. Labelling these errors, and drawing special attention to them, has diagnostic value, since we become more likely to recognize similar mistakes in the future. If we are familiar with the informal fallacy known as *argument to moderation*, for example, we will notice that kind of argument style when we confront examples, and will then respond accordingly. We'll consider several examples of informal fallacies later in the chapter.

A perennial difficulty with evaluating the arguments we encounter in daily life, however, is that we must *interpret* what we believe the author of a given argument intends. Our spoken languages are vague and ambiguous, and often subject to reasonable, competing interpretations – sometimes with amusing consequences. While I can't vouch for their authenticity, it is claimed that newspapers have run headlines announcing *Stolen Painting Found by Tree Two Sisters Reunited after 18 Years at Checkout Counter*, and *Grandmother of Eight Makes Hole in One*. Beyond ambiguities, most arguments that we confront in real life contain hidden assumptions: the author is arguing for a given conclusion but his argument includes assumptions that are not made explicit. We are forced to make an educated guess as to what hidden assumptions were in the mind of the author, as well as what was meant by the words that are used, since the plausibility of the argument may depend heavily on each.

All these difficulties can be overcome by creating a new language, a language which leaves no room for ambiguity and where every step in an argument must be spelled out explicitly. Insofar as we can translate between our natural, spoken language and our new, formal, unambiguous language, we can still hope to learn something about argument formation and evaluation. In the next section we will review the fundamentals of sentential logic as an important illustration of this method.

Arguments, as we noted in Chapter 1, are not to be understood as heated, angry exchanges between warring factions or individuals, but as consisting of a conclusion and a set of premises. The latter provide the evidence, or reasons, that are being advanced in support of the conclusion. Our concern is to *evaluate* arguments, to get better at distinguishing those arguments that lend credence to their conclusions from those arguments that should do nothing to change our opinions about the issues under discussion. To start evaluating arguments we first need to be able to identify them.

Typically arguments are not presented in neatly numbered form, with conclusions separated out. They are buried within prose, or speeches, or dialogue, but let's start with a clearer case:

1. *Onychonycteris finneyi* is a primitive bat species which dates to about 50 million years ago.
2. A well-preserved fossil of this species demonstrates, among other things, a short wing.
3. The ears of the fossil appear too small to support echolocation.
4. Therefore, this species probably relied on vision for hunting.
5. Therefore, the fossil supports the *flight first* model of bat evolution.[2]

The first three statements function as premises, describing the evidence, or reasons, that are offered in support of the conclusions. Both conclusions are introduced with the word *therefore*. There are a variety of words and phrases that help us pull out arguments from speeches and continuous prose. For example, *thus*, *so*, *hence*, *which implies* and *consequently* very often indicate that we are about to be presented with the conclusion of some argument. Likewise, the premises of an argument are sometimes identifiable through the use of words and phrases like *given that*, *since*, *because* and *assuming that*. Ultimately, however, there is no substitute for reading or listening carefully, and thinking hard about what claim someone seems to be advancing, and what reasons she is offering in support. On those occasions when we sense that others are trying to convince us of some claim we need to identify what argument, if any, is being presented. We shouldn't mistake the *opinions* of others, no matter how articulately or passionately they are presented, with an argument in defence of those opinions. If we can't identify the argument, then we must register that fact.

As we noted in Chapter 2, arguments can fail for two important, but entirely independent, reasons. First, perhaps most obviously, the premises might themselves seem implausible or suspicious. Of the above example we might wonder whether ear size is a good guide to echolocation capability, or whether the fossil was sufficiently well preserved to provide evidence of wing length or ear size. Most of the arguments we confront appeal to evidence or reasons which we might, in turn, have questions about. It is important to evaluate the premises, but arguments can also fail for a quite different reason. Sometimes the most credible, reasonable premises fail to convince us of a given conclusion (and should fail) because they simply don't stand in the right relationship to that conclusion. To better understand this idea, we turn now to the subject of *sentential* or *propositional* logic, and the kind of formal, artificial language we alluded to above.

[2] The argument is inspired by Simmons et al. (2008).

Sentential logic

The building blocks of our sentential logic are simple, declarative sentences. A declarative sentence makes a claim about the world. Declarative sentences are not questions, commands, promises or requests. Here are a few examples of simple, declarative sentences in English:

Mark is a talented football player.
There are over 450 species of shark.
Toby spends most of his spare time constructing model lighthouses from odds and ends that he finds lying around the house.
Saint Swithun's Day falls on July 15.
At the end of every rainbow there is a pot of gold.

Within our formal language we denote simple declarative sentences with italicized lower-case letters, p, q, r and so on. We might use p to denote *Mark is a talented football player*, for example. Our formal language prescribes that every sentence is either true or false, a property of the language known as *bivalence*.

Within the English language simple sentences can be embedded within more complicated ones. For example, we preface declarative sentences, as in:

I believe that *Mark is a talented football player.*
It is likely that *there are almost 450 species of shark.*
I seem to remember that *Saint Swithun's Day falls on July 15.*
It is not the case that *at the end of every rainbow there is a pot of gold.*

We can also build more complicated sentences by connecting more than one simple, declarative sentence.

If *Mark is a talented football player*, then *we should invite him to play for our team.*
Mark is a talented football player, but *he is unwilling to listen to his coaches.*

There are a great many ways to create more complicated sentences within natural languages. Within our formal system, we restrict ourselves to a handful. First, we can *negate* a sentence. Second, given two sentences, we can relate them through various connectives: *conjunction*, *disjunction* or *material conditional*. Constructing new sentences using just simple sentences and some number of these connectives yields a compound sentence. The truth-value (whether the sentence is true or false) of a compound sentence is fully stipulated within the formal language. Many of the stipulations of the formal language are exactly as we would expect. For example, within the formal language, if p is true, then its negation (often written $\neg p$, and read *not* p) is false; if p is false, then $\neg p$ is true. We here formalize how the word *not* behaves in English. If we agree that the sentence *Mark is a talented football player* is true, then we will also agree that the sentence *It is not the case that Mark is a talented football player* is false. We make the information about negation more accessible by constructing a truth-table:

Table 6.1

p	$\neg p$
T	F
F	T

The left column provides the truth values for p and the right column the corresponding value for *not p*. The truth-tables for conjunction (which we symbolize with a dot •), disjunction (∨) and the material conditional (⊃) are as follows:

Table 6.2

p	q	$p \bullet q$	$p \vee q$	$p \supset q$
T	T	T	T	T
T	F	F	T	F
F	T	F	T	T
F	F	F	F	T

As noted, we stipulate that every sentence is either true or false. If we're dealing with a pair of sentences, then within our formal language there are four combinations for us to consider: either both sentences are true, or the first is true and the second false, or the first is false and the second is true, or both sentences are false. We present all four combinations in the two left-most columns. These are the base columns. The compound sentences that we want to evaluate are listed across the top row. Looking down the truth values for conjunction (•) we see that the conjunction is true if both p and q are true but false otherwise. These assignments align with our judgements of how the and-connective works within the English language. The conjunctive sentence *Albert is tall and Beryl is old* is true if the sentences *Albert is tall* and *Beryl is old* are both true; if either of these simple sentences is false, however, then the conjunction is false. A truth-table provides the truth values for simple or compound sentences, given any assignment of values to the simple sentences.

For the most part, the truth-table for disjunctions also agrees with our natural use of the or-connective. In English, if the sentence *Albert is a doctor* is false, and the sentence *Albert is a nurse* is false, then the sentence *Either Albert is a doctor or Albert is a nurse* is false. If exactly one of these simple sentences (or disjuncts) is true, then the disjunction is true. Matters are slightly less straightforward for disjunctions when both disjuncts are true. To illustrate, suppose we are discussing Pierce Brosnan and someone remarks, *Either Pierce Brosnan holds Irish citizenship or American citizenship*. As it happens, Mr Brosnan has dual citizenship; it is true that *Pierce Brosnan holds Irish citizenship* and true that *Pierce Brosnan holds American citizenship*. Given that both disjuncts are true, what should we say about the disjunction? If we feel some uncertainty here, it's likely because disjunctions in English can be ambiguous. Sometimes such sentences are most naturally interpreted *exclusively* and sometimes they're most naturally interpreted *inclusively*. Exclusive-or sentences are true if *exactly* one of the disjuncts is true; inclusive-or sentences are true if *at least* one disjunct is true. Within our formal system, we don't tolerate any ambiguity or confusion, so we stipulate that disjunctions are true even when both disjuncts are true; that is, we adopt an inclusive-or for our language.

The material conditional ($p \supset q$) is often translated into English as *if p, then q*, although this is universally acknowledged as an imperfect translation. (The part of a material conditional that precedes the horseshoe ($\supset$) is known as the antecedent, while that which follows the horseshoe is the consequent.) In some instances the translation works well. For example, suppose I assure someone, *if you study for the exam, then you will pass*. Now suppose they study and suppose they pass. It appears I have said something true. However, if they study but fail the exam, I have clearly uttered something false. These two cases help illustrate the first two lines of the above truth-table. The third and fourth rows stipulate that a conditional statement is true whenever the antecedent is false, but this gives rise to strange results. For example, these sentences are all true, according to our formal language:

If whales are fish, then Pierce Brosnan is Irish
If New York is in Canada, then Canada is in New York
If all numbers are even, then some numbers are odd

In at least some respects, therefore, the material conditional in our formal language behaves quite unlike if-then statements in English. Nevertheless, the overall logical system that we're introducing has proved very attractive to logicians, and the material conditional plays a central role within that system. By adopting the truth-tables above, we gain a useful logical system, even if there are some difficulties translating certain aspects into English.

Now that we can determine the truth values of sentences, we can start evaluating arguments. In Chapter 2, we defined a deductively valid argument as one for which if the premises are true then the conclusion must be true. We can make this definition more rigorous by defining a valid argument as one which has a valid argument *form*, and defining a valid argument form as one that has no *counterexamples*. We define a counterexample to an argument form as an instance of that form which has all true premises and a false conclusion. For example, the argument form known as *modus ponens* has a material conditional as its primary premise, the *antecedent* of that conditional as the secondary premise, and the *consequent* of the conditional as its conclusion. The truth-table looks like Table 6.3 (below):

Table 6.3

p	q	$p \supset q$	p	$\therefore q$
T	T	T	T	T
T	F	F	T	F
F	T	T	F	T
F	F	T	F	F

We use the same base columns as before. The truth values for the material conditional have already been introduced. The two remaining columns just reiterate the appropriate base column. If we now survey each of the four rows, we observe that in no instance are the two premises both true and the conclusion false. Given the stipulations of this formal language, we can now see that it is impossible for all the premises to be true and the conclusion false. In other words, there are no counterexamples to *modus ponens*, which tells us that *modus ponens* is a valid argument form. Consequently, any argument that has the form *modus ponens* is a valid argument. Here are some examples:

1. If Mary studied for the exam, she will pass. 2. Mary studied for the exam. Therefore, she will pass.	1. If it is a nice day, then I will cycle home. 2. It is a nice day. Therefore, I will cycle home.
1. If the coffee is decaf, then it won't keep you up all night. 2. The coffee is decaf. Therefore, it won't keep you up all night.	1. If whales are fish, then I am a monkey's uncle 2. Whales are fish. Therefore, I am a monkey's uncle.

These arguments are all valid because they're examples of a valid argument form. We know the form is valid because we constructed a truth table and observed that there are no counterexamples to the form. Whether their premises are actually true is irrelevant for purposes of assessing validity.

Just as we can use the method to identify valid argument forms, so we can identify invalid forms. For example, *denying the antecedent* is known as a formal fallacy, because it is an argument form that is invalid within sentential logic. The truth-table for the argument form is as follows in Table 6.4:

Table 6.4

p	q	$p \supset q$	$\neg p$	$\therefore \neg q$
T	T	T	F	F
T	F	F	F	T
F	T	T	T	F
F	F	T	T	T

As with *modus ponens* the primary premise is a conditional statement. The secondary statement denies the antecedent (hence the name of the fallacy), and the conclusion denies the consequent of the conditional. If we look at the third row we see that both premises are true and the conclusion is false. The third row provides a counterexample to the form, and this tells us that the form is invalid. Any argument that has this form is an invalid argument. Consider the following argument:

1. If you make an *A* in this course, then you will pass the course.
2. You didn't make an *A* in this course.
   Therefore, you didn't pass the course.

The argument has the form *denying the antecedent*, which already tells us that it is invalid, but let's approach the argument in another way. As we all know, most college courses can be passed without achieving an *A*. Thus in many instances both premises are true, but the conclusion is false. Just think of the last course you passed with a *B*+, or *C*–, for example. On that occasion the two premises of our argument were both true, but the conclusion was false. Notice, furthermore that it was also false that *You made an A in the course*, and true that *You passed the course*. This assignment corresponds with the third row of our table (*p* denotes *You made an A in the course* and *q* denotes *You passed the course*), which (not surprisingly) is the row that provides our counterexample. So our argument illustrates the counterexample, but *any* argument of this form is invalid and not just those that have a false antecedent and true consequent. *Denying the antecedent* commits a formal, logical fallacy. Formal fallacies are either committed or avoided. Judging whether an *informal* fallacy has been committed can be less straightforward, as we'll see later in the chapter.

By distinguishing the *form* of an argument from *instances* of that form we can better appreciate the difference between evaluating the premises and evaluating the extent to which the premises, regardless of their own plausibility, lend credence to the conclusions. Instances of *modus ponens* are all valid arguments. Instances of *denying the antecedent* are invalid in virtue of their form alone. We don't need to attend to the specific content of these arguments to assess their validity, only their form.

Non-deductive logic

Deductive arguments are either valid or invalid, with nothing in between, but many of the arguments we confront don't have this property. My observations about a particular sports team might add some credibility to my claim that they will win their next game, but even supposing my observations are all correct doesn't guarantee a correct conclusion.

Probability theory is of enormous value and huge significance to many scientific disciplines. It's perhaps not surprising therefore that many arguments can be evaluated using probabilities. Suppose we know that a standard die is fair. We can infer the probabilities of events concerning throws of that die. The probability that on a particular throw the die lands four or higher is 1/2. If I roll the die twice, then the probability that the summed value is twelve, is 1/36. Modern statistics has developed far beyond the realms of games of chance, such that competency in the statistical methods and modelling of contemporary scientific disciplines can require significant investments of time and energy. As we noted in Chapter 5, some cognitive biases concern problems with our statistical reasoning.

Some arguments lend credence to their conclusions in ways that translate less straightforwardly into the language of probabilities. Darwin described biogeographic distributions, features of the fossil record and the results of artificial breeding programs, and also appealed to plausible ideas such as competition over available resources. He was able to advance a compelling argument, but we might hesitate to attach a probability to his conclusions. Some commentators have sought to employ probabilistic reasoning even here. The leading alternative, for analysing arguments that don't establish their conclusions with certainty, appeals to *explanatory* considerations, rather than probabilities.

As we noted in Chapter 2, scientific theories don't just describe how things are but offer explanations for why they are that way. As we also noted, one means of evaluating scientific theories focuses attention on how *well* they explain certain phenomena. Perhaps arguments like Darwin's are admired, in part, because they are perceived to explain very well those observations and phenomena that are offered in support of their principal conclusions. Explanatory considerations certainly play a role in assessing scientific conclusions. What remains to be seen is how well philosophers and scientists can specify the criteria by which we should evaluate competing explanations, how general those criteria are and what exactly they tell us about the hypothesis or theory that provides the best explanation. Being advised that we should prefer those theories that best explain our observations is of little assistance if it remains unclear how we are to evaluate which explanations are best.

Despite the large, open questions surrounding the explanatory strategy for hypothesis evaluation, it is often worth searching for ways of explaining observations via principles and assumptions that are more plausible. To foreshadow some of the themes we will visit in the remaining chapters, consider some of the examples in Table 6.5:

Table 6.5

Observation:	certain authors are unable to get their work published in reputable, peer-reviewed journals.
Explanation 1:	members of that particular scientific discipline are prejudiced, biased and unwilling to entertain conclusions that depart from the consensus view; hence they obstruct the publication of such ideas.
Explanation 2:	the work of these authors is regarded by those qualified to evaluate as critically flawed in one or more respects, in terms of methods, analysis or argument.

Table 6.5

Observation:	despite some familiarity with the scientific evidence and arguments surrounding a particular scientific claim, I find it hard to accept.
Explanation 1:	despite far greater familiarity with the relevant science, thousands of peer-reviewed articles that we're unaware of and decades of research culminating in today's scientific theories, scientists have rushed to judgement and accepted certain ideas that are not well-supported by available evidence.
Explanation 2:	I am overestimating my grasp of the relevant science, and probably unaware of vast quantities of carefully gathered and meticulously analysed empirical data; I am likely subconsciously encouraged in my deep suspicion of the claim by the frequency with which members of my community express their reservations, and its presence in the media; I am likely further encouraged subconsciously by some combination of desire that the science be wrong, reluctance to modify beliefs I've already formed and my ability to recall objections more easily than positive arguments.
Observation:	within society there persists the idea that a given scientific community is divided over an issue of social significance.
Explanation 1:	the relevant scientific community is genuinely divided, and public perception reflects this fact .
Explanation 2:	the relevant scientific community is in broad agreement about these issues, but special interest groups have successfully hidden this fact just as the tobacco industry did in the 1960s.
Observation:	scientists proclaim certain conclusions, but I can't conceive of any experiments or methods that could even potentially produce supportive evidence for them.
Explanation 1:	scientists don't actually have any evidence but accept them nonetheless.
Explanation 2:	scientists have devised methods and techniques that produce reliable evidence but which I have not been introduced to or did not properly understand.

We needn't suppose that one of the competing explanations for each of the associated observations is always the correct explanation. Perhaps institutional prejudice does sometimes interfere with publication success, even if typically individuals' inability to publish their research is an indication that the work is of poorer quality. We needn't suppose that on some occasions there aren't better explanations than those described. Finally, we need not suppose that the best explanations all share in common some particular feature or set of features. Perhaps the best we can do is to evaluate these case by case, and further explanations, on those occasions when observations such as these are being alluded to. What we are seeking is the explanation that is most reasonable. Crucially, however, what we must not do is blindly accept explanations without pausing to think hard about the possibility that there are better explanations which are not being aired.

A baker's dozen of informal fallacies

Our sketch of the fundamentals of sentential logic is a useful illustration of how logic and reasoning can be developed. The possibility that there are better explanations for our observations and that responsible thinkers will actively seek them out are central to improved critical thinking. For purposes of thinking more clearly, it is also instructive to examine some of the most common ways that arguments can go wrong. Certain errors in our reasoning are made sufficiently often that they have been named. Naming them, and describing their most salient features, improves our chances of noticing those errors when they're committed. The particular kinds of mistake we're interested are known as informal fallacies. Here are thirteen.

1. Appeal to ignorance

Arguments from ignorance draw attention either to the *absence* of evidence for or against some proposition, or some residual *uncertainty* surrounding that proposition. When such arguments are concerned with the absence of evidence *for* some proposition, they conclude the proposition is false; when such arguments are concerned with the absence of evidence *against* some proposition, they conclude that the proposition is true. The absence of good evidence supporting a particular hypothesis about the origins of life on Earth, for example, might be offered as reason to dismiss that hypothesis as false. Uncertainties surrounding the safety of a particular medical procedure might be offered by someone as a reason to suppose that the procedure is unsafe. Such inferences may be judged fallacious because the absence of evidence is sometimes attributable to factors other than the truth or falsity of the proposition we're most concerned with. Evidence in support of some hypotheses is extremely difficult to obtain. It might require technological or methodological breakthroughs, or the good fortune that the right person may happen to look in the right place at the right time. The uncertainties that surround much of our understanding and practices, moreover, are entirely consistent with having a large quantity of good evidence that shouldn't be overshadowed by perhaps small, remaining doubts. Thus the absence of certain kinds of evidence isn't always a reliable indicator that our target hypothesis is false, and the absence of certainty shouldn't distract us from evidence that might nevertheless be important and compelling.

It would be a mistake, however, to suppose that the absence of evidence is never instructive. To take a rather trivial example, if there was an elephant in the room, then there would be clear, compelling and unavoidable evidence for the animal's presence.

This fact, coupled with the absence of any evidence of an elephant, is a good reason for concluding that there is no elephant in the room. Importantly, the strength of this argument proceeds not from the absence of evidence alone, but that absence in combination with information about the ease with which such evidence could be gathered if it was available. More seriously, drugs can be subjected to rigorous testing. The rigour of those tests might be sufficient to conclude that, given the absence of evidence that the drugs have dangerous side-effects, it is reasonable to regard the drugs as safe. The absence of evidence of harm might, in these circumstances, provide good reason to suppose that the drugs are not unsafe. What distinguishes a fallacious argument from ignorance from a felicitous argument based on the absence of evidence is the confidence with which we can reasonably attest that our efforts to obtain evidence for some claim would have discovered such evidence if it existed, and hence that the absence of evidence justifies our assessment.

2. Argument from incredulity

Science presents us with some extraordinary claims. Some of those claims have become so entrenched that we don't even notice how incredible they are. Galileo's contemporaries were incredulous when he argued that the Earth orbits the Sun and spins on its axis. These sentiments are understandable, but remind us that just because we find something hard to accept is itself a poor reason for rejecting it. Nevertheless, in many cases conclusions are resisted because we find them too remarkable, too incredible, too outlandish. Incredulity can attach to certain *conclusions* about the world, and lead many to reject those conclusions, but also to certain *processes*. Of certain biological features it seems all too obvious that they *must* have been designed; the idea that they could have arisen from an unguided process appears absurd. Yet evolutionary biologists are able to explain in remarkable and growing detail just how certain features arose through a process of Darwinian natural selection. We will have more to say about this topic in Chapter 9.

3. Appeal to false authority

For a variety of reasons the opinions of some individuals are given more attention than the opinions of others. Celebrities are regularly interviewed on radio and television, in newspapers, magazines and websites. Politicians share their views on a wide variety of subjects, views that are often reported by journalists. The fact that some people's opinions are more likely to be *heard*, however, is a poor reason for supposing that those opinions are more likely to be accurate. Michael Crichton was a novelist, who authored *Jurassic Park*, among other best-sellers. Crichton described anthropogenic climate change as a hoax. Jenny McCarthy is a model and television presenter who has repeatedly suggested that childhood vaccinations increase the chances of developing autism. President George W. Bush announced in 2005 that he was in favour of teaching the controversy surrounding Intelligent Design theory. In all these cases, our present concern is not whether Crichton, McCarthy or Bush is making sensible, reasonable judgements, but whether *their* opinions should carry special weight or significance.

Crichton is not a climate scientist, McCarthy has no medical training and Bush has no expertise to evaluate the merits of Intelligent Design. They haven't spent years familiarizing themselves with, developing, amending and evaluating the relevant methods, techniques, theories and models. When we appeal to the opinions of those who have achieved some degree of fame as evidence for conclusions they have no relevant expertise to evaluate, we are guilty of *appealing to false authority*. Crichton, McCarthy and Bush have their respective opinions. None has the relevant qualifications for us to defer to their judgements on these issues. It would be a mistake to suppose that Crichton's or McCarthy's or Bush's opinions carry equal weight with those who do have relevant expertise, much less that their opinions could come close to outweighing the consensus opinion of an entire field of inquiry.

Importantly, what's under discussion here is not whether individuals are capable of formulating interesting and important objections to prevailing theories and ideas, but whether their status as authorities in unrelated fields is itself a reason to value their opinions more highly than any others. Second, such arguments are fallacious when they appeal to false or *unqualified* authorities. The restriction does not entail that all arguments employed by qualified individuals are sensible and persuasive, only that appealing to the conclusions of those individuals' arguments is not an example of fallaciously appealing to false authority. This second caveat does, however, raise a further complication, for judging whether an individual is appropriately qualified is an issue over which reasonable people can disagree. In discussions of climate change, for example, the opinions of an atmospheric physicist might seem more germane than those of a quantum physicist, but the opinions of the latter might be judged of greater value than a layperson. Whether appealing to the opinions of the quantum physicist constitutes an appeal to false authority is perhaps contentious. More generally, informal fallacies always permit of borderline cases, where it might be unclear whether a particular fallacy is being committed or not.

4. *False choice*

Suppose we know that Paula suffers from an acute phobia. If we reason that Paula is afraid either of snakes or spiders, and then establish that she is not afraid of snakes, we will conclude that Paula is afraid of spiders. However, our conclusion is reasonable only if Paula's fear really does concern either snakes or spiders. If we know only that Paula has a phobia, then the fact she's not afraid of snakes is entirely consistent with her being afraid of heights, water, dogs or the number thirteen. More generally, when we are presented with a list of alternative explanations for some phenomenon, and are then persuaded that all but one of those explanations is unsatisfactory, we should pause to reflect. Before conceding that the remaining explanation is the correct one, consider whether other plausible options are being ignored or overlooked. The fallacy of false choice misleads when we're insufficiently attentive to an important hidden assumption, that the choices which have been made explicit exhaust the sensible alternatives.

5. *Appeal to false cause*

False authorities and false choices might each generate faulty conclusions, as can appeals to false *causes*. The fallacy of appealing to a false cause occurs when we fail to distinguish correlations from causation. There might be more obesity in developed countries, and there might be more bottled water as well, but no-one would suggest that the answer to the obesity epidemic is to ban bottled water. Observing a correlation between variables does not establish that either causes the other. This doesn't stop people from observing that a given disease and a strange new technology are each becoming more prevalent, then inferring that the latter must cause the former.

6. *Appeal to the masses*

'Three men makes a tiger' is an ancient Chinese proverb, derived from a story of a king who, when questioned, admitted that he would not believe that a tiger was in the marketplace on the basis of one man's testimony alone, and would be suspicious even if two men testified to having witnessed a tiger in the marketplace. Adding the testimony of a third man would, conceded the king, be sufficient to convince him. The story illustrates a further, important and seductive kind of fallacy, variously known as *appeal to the masses*, *bandwagon fallacy* or – the fancy Latin name – *argumentum ad populum*. Arguments that commit this mistake appeal to the fact that a belief is widely held, offering its popularity as evidence that the belief is correct (or probably correct). The problem is that those who are in the majority are not always right. The fact that a lot of people believe something to be the case is not always a good reason for assenting to the same.

7. *Straw man*

Within any debate, participants must present aspects of the theory or conclusion they are suspicious of, before offering criticism or objections. Presenting the ideas of others in a manner that is fair and accurate, however, is not always easy. Sometimes ideas are attributed to others which they do not hold, but which appear similar enough to views they do hold that we might not notice the difference. If the falsely attributed ideas are easy to refute and even ridicule, we might be dealing with an instance of the straw man fallacy. We saw an example of this fallacy in the Introduction, in our discussion of the tobacco industry. Industry members were keen on drawing attention to individuals who smoked heavily throughout their lives, seemingly without suffering any ill effects, as well as individuals who developed lung cancer despite never, or rarely, having smoked. These were offered as evidence that cigarettes couldn't cause lung cancer. The objection only succeeds, however, if health professionals had argued that all and only smokers would develop lung cancer, but this is a gross distortion of considered medical opinion. That some smokers avoid lung cancers and non-smokers develop them is *expected*, on the assumption that smoking merely increases the chances of developing lung cancer. The appearance of a serious objection only emerges if we overlook the fact that the tobacco industry was here attacking a straw man.

8. Red herring

As conversations and debates unfold a variety of considerations may be introduced. If we're not vigilant, then we may find ourselves distracted by ideas that seem relevant to the issue at hand but are quite irrelevant, on closer inspection. The financial costs of proposed solutions to social problems will be entirely relevant to certain conversations, but if we are attempting to evaluate the safety concerns within those solutions, or the likelihood that they'll succeed, or the side-effects of pursuing such solutions, then the financial costs will be largely irrelevant. It is important to keep the object of our discussion in focus.

9. Hasty generalization

Most of us are aware that if a coin lands heads three times in a row this is not good evidence that the coin is biased. We know that a bigger sample is needed before such judgements are justified. However, in many circumstances we do demonstrate a greater tendency towards supposing that our experiences are reliable guides to more general claims. We suppose that our personal experience of the weather is relevant to evaluating the possibility of global warming. We suppose that one story about a friend who claims to have suffered an adverse effect to a medical procedure can help inform our evaluation of the procedure's safety. When we infer facts about a population on the basis of insufficient evidence, because the sample is too small or is unrepresentative of the broader population, we are guilty of committing a hasty generalization.

10. Appeal to motives

Conclusions are sometimes dismissed on the grounds that their advocates have an independent motive to defend that conclusion. If we learn that the tests for a particular drug were funded and overseen by the pharmaceutical company that stands to profit from its sales and distribution, we may feel inclined to distrust the results of those tests. However, such reactions may prompt concerns that we are committing the fallacy of *appealing to motives*. It is quite conceivable, of course, that research scientists who are funded by global corporations will justifiably reach conclusions that are convenient for the industry or corporation that funds them. Research teams funded by large agricultural companies might learn that newly developed crops are more resistant to disease, more nutritious or more tolerant of variable weather patterns. If there are problems with the research, whether it be in the design, methods, analysis or conclusions, then these may be important. However, such problems concern the details of the studies not the origin of the funding. Dismissing research on account of the motives of those conducting it is problematic because having independent reason for wanting to find evidence for some conclusion is consistent with finding good evidence for that conclusion. We should be aware of the appeal to motives fallacy, but in Chapter 7 I'll suggest that evidence of independent motives is information that we shouldn't be too quick to ignore.

11. Appeal to possibility

In response to arguments offered for scientific conclusions, people will sometimes draw attention to an alternative conclusion that *could* be the case. In some instances, the alternative hypothesis is defended by appealing to additional evidence and considerations. But not always. Sometimes the mere possibility that something *could* be the case is advanced as an objection. As a defence of the alternative conclusion, this is a strange strategy. Bertrand Russell observed that it *could* be the case that God made the world five minutes ago, complete with heaps of evidence that the world was much older. Part of the evidence would be creating in all of us false memories. We *seem* to remember events that happened yesterday, but perhaps these apparent memories are all illusory. Similarly, it *could* be the case that someone broke into my house last night, stole all my possessions, replaced them all with identical replicas and left without leaving any evidence of their mischief. I suspect the police would be less than impressed if I filed such a complaint. The mere possibility that something happened a particular way is a poor reason for thinking that it did happen that way, or that it probably happened that way or even that there's a small (but non-negligible) chance that it happened that way. When people are in search of justification for a particular idea the mere possibility that things *could* be a particular way is not a good defence.

12. Argument to moderation

Arguments to moderation assume that the truth must lie somewhere between the positions being defended by opposing sides. If vitamin manufacturers suggest that daily supplements are beneficial, but others claim they have no benefit, then an argument to moderation concludes that the benefits are real but not as significant as the manufacturers assert. If advocates insist that nuclear power is extremely safe but critics respond that nuclear power is very unsafe, then moderation suggests the true risk must lie somewhere in the middle. Sometimes arguments to moderation are no doubt the correct response to diverging conclusions. However, it is a style of argument that can also lead us astray. If we're told by one group that the evidence for some conclusion is very compelling, but a second group insists that the evidence is of extremely dubious value, it will be tempting to suppose that the actual value lies somewhere in the middle of these two assessments. If those on one side of the issue are highly qualified, highly skilled and extremely capable of evaluating the evidence, while those on the opposing side harbour an independent agenda that requires them to deny the evidence and have few relevant credentials, then arguing to moderation is both fallacious and potentially dangerous.

13. Everyone's entitled to their own opinion

Many more informal fallacies have been identified, analysed and discussed than those we've introduced here. One more fallacy is worth briefly discussing. It is perhaps best understood as a version of the *red herring* fallacy, but is repeated with sufficient frequency that it deserves special attention. When discussing the merits of alternative points of view, it is not uncommon for someone to declare, perhaps somewhat defensively, that everyone is entitled to their own opinion. Jamie Whyte has a nice discussion of this strange form of reasoning.[3] He argues, first, that it's not entirely obvious that we *do* have such a right. Second, the idea that we are each entitled to our own opinion is almost always irrelevant to the context in which the claim is introduced.

Within the context of a conversation about the merits of competing ideas and opinions, Whyte suggests that appealing to the right to one's own opinion is best interpreted as placing a requirement on others to desist from trying to change that individual's views. But, as Whyte suggests, if people hold an opinion that places them in mortal danger then we would seem to have a responsibility to rid them of the dangerous attitude. Even ignoring practical matters, and granting people their right to hold opinions that are poorly supported by available evidence, it's rather strange that people would want to hold an opinion simply because they have a right to. To illustrate the charge of irrelevancy, suppose we're engaged in a discussion about the merits of teaching comprehensive sex education courses in high schools, as opposed to the abstinence-only programs that some individuals prefer. Arguments, ideas and objections flow back and forth. At some point, however, some people simply announce that – despite all that's been said – they're entitled to their opinion. How is this relevant? The point of such discussions is to try and establish which ideas are best

[3] Whyte (2004).

supported by the evidence and arguments we have available. People's right to their own opinion fails to advance the discussion. Either an opinion is held for good reasons, in which case there is no need to defend the opinion by appealing to rights, or an opinion is held despite the absence of good reasons, in which case that opinion clearly should influence neither behaviour nor decisions nor the development of social policies. Either way, in almost every context, having a right to an opinion is irrelevant.

Conclusions

Identifying exactly what people are arguing for and the reasons they're advancing in support, recognizing alternative ways of accounting for available evidence and evaluating the merits of these competing conclusions are all, often, difficult, complex, demanding tasks. Responsible reasoning and argument evaluation require considerable effort. Becoming sufficiently well informed about issues that matter to us requires significant investments of time and energy. Such efforts are necessary, however, if we are to achieve reliable beliefs, make informed decisions and improve the chances of attaining our goals. The printed word might be wrong. The proclamations of charismatic, confident people who talk fast, use jargon that might be unfamiliar to us and claim expertise over certain domains might be wrong. People announce *facts* that are highly dubious. People appeal to false authority, the masses, emotion, motives or ignorance. People shift from one defence of their opinion to another, misrepresent the ideas they're challenging or introduce considerations that aren't strictly relevant. In this chapter, we have surveyed ideas that can help cultivate better reasoning, but there are few helpful short-cuts to improved critical thinking. We must instead continue to ask ourselves: What idea, or conclusion, is someone trying to convince us of? What reasons or evidence are they offering in favour of the conclusion? How are those reasons intended to support the conclusion? How much support would they provide for the conclusion if we suppose they're true? How plausible are the premises? Are there better ways of gathering relevant evidence? Can we think of alternative ways to account for the conclusion? Are these alternatives better or worse?

Discussion questions

1. It is sometimes suggested that people can *choose* to believe whatever they *want* to believe. Do people really have this ability? (For example, can you choose to believe that you can fly?) If there are limits to what we can choose to believe, what do these limits tell us?
2. Why should we care to improve our critical thinking skills?
3. What are the differences between good critical thinking and good debating skills?

4. Find an online publication. Read some articles and the responses from readers. What fallacies of reasoning can you identify?
5. Can you come up with examples of the informal fallacies described above? Discuss your examples with others.
6. Do you think our understanding of why informal fallacies are committed might be illuminated by work in cognitive psychology? Are there any connections you can already sketch?
7. People differ in terms of mathematical proficiency, reading proficiency, athletic ability and so on. Improving these capacities is largely a result of instruction and practice. Is the same true of critical thinking? Does society place sufficient emphasis on improving critical thinking? Do we receive enough instruction? Are there costs associated with devaluing basic critical thinking skills?

Suggested reading

There are very many texts on logic and critical thinking. Hurley (2014) is very comprehensive for those interested in improving their logic, reasoning and critical thinking skills. Salmon (2006) is more concise, but very thorough. Schick and Vaughn (2013) adopt a less formal approach.

7

Created controversies and how to detect them

Scientists often disagree. As we saw in Chapter 4, dissent may be reasonable, appropriate and even desirable. If competing approaches to questions of mutual interest each enjoy some measure of empirical success, then we might hope that they would all receive continued attention, approximately commensurate with their perceived degrees of success and promise. Reasonable scientific disagreement can help clarify the topics under dispute and may function as a stimulus for new research and a conduit for progress. Sciences might not technically require controversy in order to advance, but they are certainly familiar bedfellows.

The details surrounding most scientific debates never reach the public ear. The finer technical points concerning how best to measure, interpret or analyse particular results may be largely incomprehensible to those who lack relevant training. Some apparent scientific controversies, by contrast, receive enormous public attention, whipping up great storms of sound and fury. Journalists lavish the issues with great attention. Popular books are written. Proponents for competing views stage public debates, appear on popular radio and television shows and oversee labyrinthine websites which link to new studies that sound authoritative and important, provide commentary that seems erudite and informed and offer criticisms of alternative scientific conclusions that appear grave and compelling. These appearances of controversy, often surrounding socially important issues, might be regarded as evidence that genuine disagreement exists within the relevant scientific community. As we know, however, sometimes appearances are deceptive.

Deceptive appearances can pave the way to significant problems. If those who are most qualified to evaluate an issue are in broad agreement about the correct conclusions to draw from available evidence, but the public perception is of experts deeply divided, then the talents and labours of our experts are in danger of being lost. Ignoring those who are most qualified to judge an issue can only be detrimental. In the final section of the book, we will look at several examples where many claim that the relevant sciences are being baselessly ignored. In this chapter, we will consider in more detail the very concept of a created, or manufactured, controversy. Among the questions we will address are the following. How and why are controversies created? How can we spot a manufactured controversy? And how should we respond to evidence that an apparent controversy is not controversial within the relevant scientific community? We should start, however, by better articulating exactly what we mean by a created controversy.

Defining the term

We have alluded, on several occasions, to the concept of a created, contrived or manufactured controversy, but have yet to properly define what we mean. Leah Ceccarelli suggests that:

> A scientific controversy is 'manufactured' in the public sphere when an arguer announces that there is an ongoing scientific debate in the technical sphere about a matter for which there is actually an overwhelming scientific consensus.[1]

A successful campaign of creating, or manufacturing, controversy may thus induce significant public confusion about the state of the sciences. A created controversy is unlike a genuine scientific controversy since the latter involves no 'overwhelming scientific consensus'. Ceccarelli proceeds to relate the motivations for manufacturing controversy to the ambitions of special interest groups. Those motivated to create or grossly exaggerate the appearance of controversy are not concerned principally to better understand some aspect of ourselves or the world, nor to investigate issues that are of general public interest or importance. Instead, such groups are motivated by profits, ideology or a desire for political influence. A corporation worries that scientific conclusions are a threat to its financial future and thus embarks on a strategy of undermining the credibility of the relevant sciences and those who defend them. A religious group regards a particular behaviour as depraved, and thus expects others to abstain. When religiously based arguments for the immorality of the act fail to persuade many members of society, and when scientists suggest that there are no dangers associated with the act, or when technologies reduce what risks are apparent, then religiously motivated groups might feel motivated to inflate doubts about the safety of the technology or the merits of the scientific studies.

We will return to the significance of motives to the concept of a created controversy, but now let's note that Ceccarelli's original definition leaves open the possibility that controversies might be created *unintentionally*. On occasions when public understanding departs radically from expert analysis, but where it may be unclear that any given group intended to create that confusion, we have no less reason to strive for reasonable judgements. Consider a group that is united by a common cause and seeks answers to a shared concern. If they stumble upon a solution that satisfies them,

[1] Ceccarelli (2011).

promote it with sufficient conviction and sincerely find scientific responses and rebuttals unpersuasive, then the effect could be a heightened state of public confusion surrounding an issue that is scientifically uncontested.

The activity and behaviour of the group has resulted in something that may share important characteristics with ideologically or financially motivated controversies, despite the absence of any intention to confuse. We should be no less concerned to lift the clouds of confusion that surround those controversies that are only inadvertently created. Furthermore, squabbles over the actual intentions of a group can become an unfortunate distraction that hinders reliable evaluation. Our priority should be to establish whether the appearance of controversy represents a genuine division within the relevant scientific community. If we place too much weight on the role of intention, we might thereby tempt the mistaken inference that the absence of evidence for pernicious intentions implies genuine disagreement within that scientific community.[2]

The most significant feature of a created controversy, as I'll use the term, is the conflict between expert and public evaluations. If the overwhelming majority of people qualified to judge a scientific issue are in broad agreement over what conclusions are justified, but those conclusions are grossly misunderstood, denied, regarded with acute suspicion and even ridiculed outside the community of experts, then we are dealing with a created controversy. A consequence of this description is that simple public misconceptions might seem to qualify. It is instructive to reflect on what distinguishes these from more paradigmatic instances of created controversies.

Contrary to popular belief, the Great Wall of China isn't really visible from space, and certainly not from the moon, the colour red is no more likely to anger bulls than any other and your fingernails and hair won't keep growing once you're dead. We could fill a book with popular beliefs that aren't actually so, but these are not our target. The primary difference between created controversies and merely popular false beliefs is twofold. First, misconceptions like these listed are of simple design, memorable and easily communicated. It is likely a consequence of such features that they propagate so successfully. Convincing someone that a certain common belief is false, furthermore, is often quite straightforward. Created controversies, by contrast, involve what can feel like an overwhelming quantity of evidence, arguments, refutations and ideas. Matters are not simple, there is too much to remember and consequently what is heard and passed on is partial, inadvertently modified and often conflicts with related ideas that are also in public circulation.

The deluge of information surrounding a created controversy results in the second major distinguishing characteristic – a collective state of confusion and uncertainty, rather than the widespread acceptance of something that is actually false. Individuals

[2] When it comes to assigning blame for social injustices, the intentions of a group are more pressing. The legal cases brought against the tobacco companies in the 1990s established that those companies deliberately sought to deceive the public about the risks of smoking cigarettes. The rulings in those cases relied centrally on this evidence.

misunderstand the import of particular observations, arguments, evidence or objections. They echo the conclusions of mainstream science, but for the wrong reasons, or adopt modified versions of scientific theories without any good reason for doing so. Perhaps many people believe that a penny dropped from the Empire State Building has the potential to kill a person. It doesn't. The public dialogue surrounding climate change, by contrast, is a buzzing, complicated, confusing cacophony. Open the floor for discussion, and people deny that the climate is changing, or admit it but then blame volcanic activity, unusual solar activity, El Niño, or just appeal to natural variability. Others propose that promoting clean energy is appropriate, whether or not the predictions of climate sciences should be trusted. The mild winters and record summers we've been experiencing are alluded to, but someone else remembers a report about snow *in Texas* . . . in June! Someone mentions Al Gore's average monthly electricity bill, but fails to explain how this is relevant. Rumours circulate that it hasn't got any warmer since 1998, or 2005, or 2006, that Antarctica is getting bigger, that polar bears are thriving. We hear that the temperature record is unreliable, that climate models are unreliable and that climate scientists are involved in a global conspiracy. We hear too much. We feel overwhelmed. We realize that *we* don't have all the answers, and thus it becomes easier just to suppose that no-one really knows, hence those that *think* they understand climate change, and predict future economic and societal hardship, are being alarmist.

The difference between popular myth and created controversy is thus the difference between a simple, false belief and a large set of ideas whose inter-relations are complicated and which consequently can leave us unsure both about what we believe and how much confidence we should have in those beliefs. Part of the explanation for our individual and collective confusion is that the sciences are sold to us as undecided. Another part of the explanation is that circumstances such as these provide very fertile ground for the kinds of cognitive bias described in Chapter 5. The availability heuristic ensures that people will be unduly influenced by evidences that are more vivid, or more memorable. Confirmation bias describes our tendency to favour ideas that fit our prior opinions and attitudes. It's beyond the scope of the book to draw any careful connections between cognitive biases and created controversies. Our priority is evaluating whether expert opinions diverge radically from public perceptions and, if so, how we should respond. We will leave it to others to sort out and understand more fully the psychological and sociological mechanisms by which such divergence arises.

According to Ceccarelli, created controversies are not in dispute *among appropriate experts*, but we have thus far said nothing of how we distinguish the expert from the charlatan. Clearly, if we gerrymander our population of experts in the right way, then nothing will qualify as a manufactured controversy since the issue under review *will* be in dispute among those we've selected as our experts. The problem therefore is how we, as laypersons and non-experts, can reliably identify the experts. We approach this problem, first, by thinking about what an expert is and, second, by considering what indicators of expertise are available to us, how accessible they are and how reliable they are. A subsidiary question is how we might informally but helpfully rank experts, since it would be naïve to suppose that all experts are equally reliable evaluators of complicated issues.

Alvin Goldman suggests that expertise has both a comparative and non-comparative connotation.[3] Experts are *more* knowledgeable than novices within their area of expertise, so hold more true beliefs and fewer false beliefs about the subject matter. Knowing more than most, however, doesn't quite seem sufficient: experts are also in possession of a substantive number of true beliefs within their area of proficiency. Expertise involves passing a certain threshold of knowledge and understanding, where that threshold is surpassed by relatively few people. Finally, beyond knowing more and knowing much, Goldman suggests that experts are able to utilize what they know to better answer questions, and more easily assimilate new information, insofar as these are pertinent to their domain of expertise. It is useful to keep these ideas in mind when reflecting on our own level of understanding and any tendency on our part to dismiss expert opinion in preference for our personal evaluations of complex topics.

Goldman's suggestions provide a plausible account for what an expert is, but as Goldman is well aware they can't help the novice distinguish experts from non-experts. If we are not positioned to evaluate which claims are true and which are false, within some technical, complex sphere of inquiry, then we can't evaluate whether others are more knowledgeable than the general populace. If we do know which claims are true and which are false, then we're experts ourselves. What is needed are indicators of expertise that even the novice has access to. Goldman suggests various means by which novices might reach more reasonable judgements concerning claims to expertise, including our expectation that experts will have the appropriate training, qualifications and experience. These, I submit, represent by far the most accessible and important means of determining expertise. For our purposes, it is reasonable to define experts in terms of their qualifications.

Of course, it is true that some who have earned medical degrees might nevertheless know less about certain health conditions than others who received no formal medical training but who have dedicated substantial time and energies to learning as much as they can. Likewise for any other academic disciplines. However, for the purposes of our *definition* we don't need to identify a group that consists of all and only the most informed individuals on a given subject. We can afford to be less precise than that. Our project is to identify whether expert opinion differs from public perception. If the criteria by which we judge expertise become too burdensome, then the project becomes unmanageable. The advantage of attending to formal qualifications is that these can be reasonably well assessed even by those who don't have time and

[3] Goldman (2001).

resources to become experts on a subject themselves. Just as important, at this point we are not concerned with the question of whether we should always adopt the expert opinion as our own. We can leave room for the possibility that non-experts may offer critical objections to the prevailing attitude of the experts, hence that these cast serious doubt over scientific consensus. As such, we needn't be overly concerned about excluding some individuals who have substantive contributions to make to a given debate.

Much of Goldman's remaining discussion is directed towards the kinds of evidence that facilitate relative evaluations of experts, and it is certainly true that more fine-grained evaluations of relative expertise are often possible and helpful. For example, across a range of issues biologists are the relevant community of experts; however, within some discussions degrees and training in molecular genetics will be less relevant than a similar level of experience and qualification in ecology and evolutionary biology. Second, scientists spend time conducting research, publishing in peer-reviewed journals and securing grants for future work. Some are more successful than others, a consideration that again seems germane for purposes of comparing the value of competing expert opinions. Goldman observes that experts may have agendas which could compromise the reliability of their judgements, a point we will return to below. The proposal I have in mind, however, is that appropriate training and qualifications establish our field of experts, for purposes of defining a created controversy. If something approaching consensus exists among this population, but public evaluations are decidedly mixed and confused, then the conditions of our definition are satisfied. In some instances, we might offer evidence concerning the relative reliability of different experts, but these will fall within the more general project of establishing the status of a given scientific issue. Having now offered a more careful discussion of what a manufactured controversy is, it's time to consider *how* they're created.

The importance of uncertainty

In the late 1960s a tobacco industry document announced that:

> Doubt is our product since it is the best means of competing with the body of fact that exists in the mind of the general public. It is also the means of establishing a controversy.[4]

As the quote suggests, spreading doubt and uncertainty was central among tobacco industry efforts to create controversy, and it is equally central to other cases. The more that uncertainty is promoted, the more people will think that the issues are genuinely unsettled, and hence the more likely they will be to decide their opinion based on emotion, desire or rhetoric. That the appearance of substantive doubt can always be

[4] Quoted in Proctor (2012).

achieved is a fact we can trace back to the problem of induction discussed in Chapter 2. We emphasized at that point that no body of evidence can ever prove, beyond all doubt, the kinds of scientific claims that we're interested in, and thus it is always conceivable that we might be mistaken in some respects or others. The history of the sciences and our contemporary attitudes towards the sciences were observed only to enforce the important idea: scientific conclusions, methods and theories are always subject to modification, retraction and refinement. Since the evidence and support we can tender in favour of some theory will always fall short of establishing its conclusions with certainty, then opponents of a scientifically well-established opinion can always insist that *available* evidence is sufficient neither to warrant belief nor significant economic investment, nor to justify changes in attitude, behaviour or policy.

The mere fact that it is always possible to query scientific ideas creates the opportunity for groups to repeatedly and strenuously draw attention to the ways in which a complex set of ideas might be mistaken or incomplete, thereby distracting us from whatever positive evidence and arguments are available. Uncertainty and doubt are magnified and inflated. Doubts and objections can be made increasingly available for public consumption, simply by making them more visible. Insofar as many people labour under the misapprehension that science progresses by accruing a steadily growing body of facts, any inadequacy with a scientific conclusion or theory will appear critical, or at least sufficient to justify suspending belief. And there are a variety of dimensions along which such uncertainties can be flagged. Groups might claim insufficient evidence surrounding the merits of the methods being utilized or the assumptions being invoked; objections might be raised that a sample is unrepresentative of a larger population or that potentially confounding variables have not been adequately accounted for. The appearance of unanswered questions and reasonable doubts will lead many to dismiss as a hoax a much broader set of ideas. If science ever established its conclusions with certainty, this attitude would perhaps be reasonable, but it doesn't. The absence of certainty, as we've noted, is entirely consistent with having extraordinary evidence supporting those conclusions, reminding us to place no significance on the mere appearance of doubt.

In some instances, an obsession with uncertainty relies purely on the idea that available evidence and arguments are not enough. More studies are requested; more data are needed, it is claimed. The beauty of this cautionary attitude, for purposes of creating controversy, is that these seem to instantiate important *scientific* virtues. What could be more paradigmatic of the meticulous, careful, prudent approach of scientists than a request for more time, a few more studies and the chance to iron out a few wrinkles? And those requests seem all the more reasonable when the alternatives can be described as requiring enormous financial investment that will stunt economic growth, or as exposing a generation of children to unnecessary and detrimental risks.

What's easy to miss, however, is that requesting more data, before any conclusion will be admitted or policy supported, is also a wonderfully effective method for postponing meaningful action. Demands for more studies can be repeated again and again, whether or not those qualified to judge the issues regard the additional studies as remotely helpful. Demanding more data might not indicate any real need, only one group's unwillingness to admit what available data ably support. The strategy moreover involves no detailed engagement with flaws in a study or line of reasoning, only a proclamation that the available evidence hasn't convinced these particular critics. The standards by which someone evaluates claims, however, can be elevated to thoroughly unreasonable heights.

Uncertainty is useful even in the absence of specifics, for those who attempt to undermine well-established scientific conclusions, but specifics are also helpful. Kuhn recognizes that theories can never explain everything we expect from them. There are always puzzles and anomalies in any discipline. Partly in consequence, within any given scientific discipline, there will be genuine controversies about particular methods, assumptions or conclusions. Phenomena that are thus far only poorly understood, or observations that are hard to fit within the prevailing framework, or genuine disputes within the scientific community are useful for eroding confidence in those scientific fields, but this ubiquitous feature of the sciences shouldn't shake public confidence. Manufacturers of controversy rely on our forgetting that all theories live alongside anomalies and puzzles. Spotlighting the puzzles, conflicting reports and anomalies that surround some scientific theory involves nibbling at the edge of a large and complicated set of issues, demonstrating that some of those edges are vulnerable to competing interpretations or explanations, then concluding that the entire theory or enterprise is in total disarray.

Finally, there's no obvious cost associated with flooding a discourse with cheap, flimsy objections to prevailing scientific ideas, and so more is better. As we noted, it is often the sheer number of ideas orbiting a subject that overwhelms the novice and results in arbitrary opinions. Part of the strategy for creating controversy, therefore, involves churning out as many objections as possible, and as many reasons to doubt the inconvenient truth, thereby increasing the chances that for someone, somewhere, one of the complaints will seem compelling. Notice, furthermore, the gross asymmetry between raising objections and defending scientific explanations. It requires little imagination, acumen or intellect to keep expressing doubts and demanding better evidence. Properly conducted research, by contrast, takes time and energy. Defending the scientific interpretation against all doubts requires nothing less than a complete theory of everything (and even then we won't have explained why there's something rather than nothing).

Mere uncertainty is neither surprising nor troubling. It is unsurprising because every scientific claim, no matter how well supported by evidence, must live alongside some residual uncertainty. Uncertainty in itself isn't troubling, because not having answers to certain questions is wholly consistent with having very compelling answers to very closely related questions. Appeals to ignorance draw attention to what is not known and then infer conclusions that are simply not warranted by the uncertainty. As noted in the Introduction, there is a lot we don't know about smoker habits and individual cases, but these uncertainties are quite consistent with the conclusion that smoking increases the chances of lung cancer. Communities of experts are likely to be far better acquainted with prevailing uncertainties than the layperson. When such communities conclude that those uncertainties are insufficient to undermine the core picture that the evidence collectively suggests, then *our* awareness of those uncertainties can't be sufficient grounds to convince us that the core picture is fundamentally misguided.

Additional strategies for creating controversy

While promoting and exaggerating uncertainty represent the most important strategies for creating controversy, several additional strategies also deserve mention. For example, controversy creators will *cherry-pick* both their data and their experts. When it is suggested that available evidence supports a particular conclusion and that the overwhelming majority of qualified individuals accept the conclusions, it is not thereby implied that *all* evidence points unambiguously in the same direction, nor that *all* scientists are persuaded. It should be irrelevant when critics draw attention to minority reports or name those scientists who reject the consensus position, but by making these studies and names more available, there is a good chance that members of the public will misjudge their actual frequency or significance. To illustrate, suppose we send twenty-five schoolchildren home with outdoor thermometers, instructions on how and where to hang them and an assignment to record the temperature twice every day, for three weeks, at specific times of the day. Now suppose that when the results are gathered and analysed we discover that twenty-three reports agree that the average temperature was highest in the third week and lowest in the second week. The remaining two reports reach conclusions that disagree with the majority and disagree with each other.

Without extraordinary evidence to the contrary, it is clearly most reasonable to suppose that the majority is correct. We might be curious about why two students reached different conclusions. Merely alluding to these reports as evidence that the third week was the not the warmest, however, is patently not a good argument. When attention is drawn to particular studies, or phenomena that appear surprising relative to mainstream scientific conclusions, we should consider that even if these studies are being presented honestly and accurately, and were conducted judiciously, they might still be inconsequential when stacked against the evidence that supports the consensus view. Critics of particular scientific conclusions must convince the relevant community that those conclusions are implausible when evaluated with respect to all available evidence, not simply that they're implausible when measured against a small proportion of the evidence that has been hand-picked specifically to undermine those conclusions.

Controversy creators can cherry-pick not just data but also experts. Identifying individuals who have the relevant academic qualifications and are willing to say what you need saying can help add false legitimacy to your position. The more qualified and the more prestigious your experts are, the better, but you can work with whatever's available. A reputable physicist's dismissal of prominent ideas within environmental science can be worth a lot, within the context of the public debate, no matter what his familiarity with the relevant sciences really amounts to. That individuals are willing to challenge the majority opinion is attributable to a variety of factors. Not all factors need be ignoble, but we clearly should be wary of adopting attitudes of a very small minority simply because they're defending conclusions that we'd prefer to be true.

Detractors of mainstream sciences do advance evidence and arguments in support of their conclusions, and these deserve attention and consideration. However, when those arguments are subjected to authoritative and critical objections, the response of those who seek to create controversy is often simply to *ignore the objections*. Thoroughly discredited arguments are repeated, presumably because many readers, listeners or viewers will be unaware that the arguments have been discredited, or won't care. If groups are concerned to influence public opinion via whatever methods work, rather than advancing reasonable conclusions in light of available evidence, then it doesn't matter whether acute problems with certain claims have been convincingly detailed. Promoting uncertainty, as we saw, can involve erecting unreasonable standards of evidence, but many people won't notice if other arguments fail to clear even the most minimal standards. Controversies can be created by adopting *double standards*. Relating your cause to public *fears and desires* is also useful for manipulating public attitudes. Within political discourse, studies are championed as *sound science* or dismissed as *junk science*, but often the use of these labels is driven entirely by political agenda rather than actual scientific merit or fault.

Objections to mainstream science can be more effective at increasing public uncertainty if they appeal to scientific papers that have appeared in published journals and that include extensive tables of data, sophisticated graphs and charts, technical jargon, links to other authors, articles and so on. If we're personally unable to see why a particular argument is flawed or misleading, we are more willing to concede the possibility that it might have some merit. Importantly, however, including technical jargon, being published, being cited and so on are not always reliable indicators that a scientific argument is sensible. Special interest groups have created their own institutions with names that suggest authority and esteem, but that are concerned solely with promoting industry interests. Likewise such groups create their own journals to ease publication of papers that coincide with their agenda but would not be deemed of publishable quality by reputable journals. Second, creating a fancy chart or spouting unfamiliar vocabulary can be achieved by any good con artist. If an argument is presented in a language we can't read, we wouldn't offer any opinion over its plausibility. We do well to remember this when faced with technical claims that we lack proficiency to evaluate: the fact that accompanying prose is in English, and the majority of words are familiar, doesn't mean that we're always well equipped to reach reliable evaluations.

Leah Ceccarelli, whose definition we introduced earlier in the chapter, draws attention to the way in which creators of controversy can turn certain scientific ideals to their own advantage. For example, science is supposed to be egalitarian and open-minded. Everyone is capable of gathering observations that could transform scientific opinions. Everyone's queries and concerns deserve attention. Every scientific idea should be available for discussion, further testing and evaluation. In the face of such ideals, even a whiff of elitism or dogmatism can be greeted with derision, and interpreted as the sciences failing by their own standards. This places scientists in an awkward dialectical position. If they remonstrate too forcefully, they appear dogmatic. Ceding too much to their detractors, by contrast, gives the impression that the objections deserve more attention than may in fact be appropriate.

It's not just individuals who can be portrayed as dogmatic. Scientific attitudes might be described as *orthodoxy*, implying that they are accepted on the basis of faith rather than reasons. If dissenting views are systematically ignored or berated, then critics may claim that the entire scientific community is dogmatic, a claim that can also be used to explain why peer-reviewed journals never publish the deviating views. What we should remember, however, is that a commitment to being open-minded does not entail listening to the same argument over and over when its flaws have been convincingly revealed. There might be a good reason why certain ideas now receive little attention from serious scientists: perhaps those qualified to evaluate the idea have recognized its deficiencies and believe they have made reasonable efforts to explain them. Critics are welcome to being *new* objections and questions, but wasting others' time won't be appreciated.

Finally, while not exactly a strategy for creating controversy, manufacturers sometimes benefit from the practice of journalistic balance and a public desire for fairness. To illustrate, suppose that all physical chemists regard the process of carbon dating as extremely reliable, but a particular geologist remains unconvinced. In the interests of balance, a journalist might interview both an advocate for the technique and the geologist. The resulting article includes the opinions of each and, consequently, could easily give rise to the impression that the opinions of the scientific community are evenly divided. Without attributing any malicious intentions to the journalist, and even allowing for the possibility that some caveat might be included – to the effect that the dissenter is in the minority – many readers will walk away with the impression that the reliability of the method is far more contentious than it is.[5]

[5] There are additional reasons to be cautious about media presentations of scientific conclusions, beyond those associated with the pursuit of balance. Media reports often confuse correlation with causation, present statistics in misleading ways, are drawn towards stories that will help attract or retain readers or offer sensational headlines for the same reasons. Supposing the media presentations of the state of scientific understanding reflects the actual state will sometimes lead us astray.

Detecting created controversies

Created controversies involve significant deceit and are potentially extremely dangerous. Lives are at stake because reliable information on important issues is distorted and masked. Given that the stakes are so high, it would be useful if there were means of sniffing out such affronts to rationality. It would be useful if there were indicators that a particular debate was created and that public attitudes are unrepresentative of the opinions of qualified experts. In what follows, we'll consider three such criteria. First, the presence of certain motivations for creating doubt should raise concerns that some participants in the debate are poorly positioned to offer impartial evaluations of the issue and thus that the debate is being fuelled by one group's *desire* that things be a particular way, rather than good reasons for believing them to be so.

Our observation that creating controversy is achieved, to a large extent, by inflating uncertainty inspires a second indicator: if the arguments being advanced by one side in a debate are slanted heavily towards undermining and challenging the views of their opponents, rather than developing and testing an independent, positive thesis, we might suspect that the appearance of controversy has been manufactured. Finally, if we have evidence that groups are more invested in engaging the public, rather than qualified experts, this is also significant. Why are the opinions of the relevant community of experts not being given utmost priority? Is it because developing credible scientific conclusions is less important than shaping public opinion and behaviour? Is it because credible experts have demonstrated significant flaws with the logic of evidence being presented by dissenters? Whatever the reason, if substantive doubts, uncertainty and objections are prevalent within public discourse, but largely absent from the professional ranks, then we are clearly dealing with a created controversy. What we should do with that information remains for us to discuss. It is first worth discussing each of these three indicators in a little more detail.

Motives

Earlier in the chapter we defined a created controversy to include cases where there's no clear intention to create confusion. Absence of relevant intentions is insufficient for us to conclude that a controversy has not been created. Nevertheless, if we have reason to suspect that special interest groups have motivation to question scientific conclusions, then this should certainly give pause. The recommendation here is not that the appearance of self-serving motives is adequate reason to dismiss those arguments that serve such agendas. Doing so commits what we described in Chapter 6 as the fallacy of appealing to motives. Arguments, evidence, analysis and conclusions should ideally be evaluated without regard for who funded the research or what the researchers hoped to find. Such assessments are always best conducted by those qualified to make such judgements, but, as we've noted, it's impractical to furnish ourselves with all the necessary skills and background knowledge that would be required to properly evaluate all aspects of these issues. Furthermore, given what we know about human psychology, it is reasonable to be more suspicious of technical arguments that are advanced by those who have financial or ideological incentives to defend the conclusions that they are actually defending.[6] Thus, if the only arguments offered in support of some thesis are being advanced by those who are financially or ideologically motivated, then this degrades the value of those arguments. The conflicts of interest that arise when individuals pursue scientific questions with the hope of discovering particular answers, coupled with the practical limitations concerning how much we can learn about abstruse, technical arguments, justifiy appealing to motives for purposes of evaluating certain issues.

Two further distinctions are worth making on the subject of motives. First, the motives of those who help create the appearance of controversy are typically distinct from the reasons they provide in defence of their opposition to the scientific community. Suspicions that certain motives are playing a role need not imply that the creators of controversy are thoroughly sensitive to the influences that are directing their efforts. More broadly, for all agents participating in debates, we should separate the reasons they articulate within discussions from explanations about why they hold those beliefs. In subsequent chapters, we will attend both to the reasons most frequently *given* for doubting or denying mainstream science, as well as speculate about some of the subconscious influences for such opposition. The distinction between sympathy for the *cause* and sympathy for the *deception* is also important. Some people are morally opposed to abortion, but others are not. Some opponents of the practice might attempt to deceive pregnant women about the risks of the procedure. Even those who have moral reservations about terminating pregnancies, however, are clearly not obliged to distort the relevant sciences and deceive people about the risks. The ethics of lying raises big questions, and some would argue that in some circumstances the ends can justify the means. I'll leave it to the reader to decide whether any of the cases we'll discuss justify widespread dishonesty and deception.

Absence of alternatives

Our second indicator that a controversy is created arises because groups that want to challenge inconvenient scientific claims don't have good scientific arguments on their side. Such groups must therefore resort largely to magnifying uncertainty and manufacturing doubt. Thus, if it appears that scientists are divided on an issue, we should consider whether one side seems to invest a disproportionate amount of time, energy and effort into undermining the theories and conclusions of their opponents, rather

[6] Bekelman et al. (2003) provide another good example of biases infecting scientific work. They reviewed financial conflicts of interest arising within biomedical research and found that where these conflicts were apparent there was a higher chance that researchers would find results that were favourable to the sponsoring body.

than promoting, developing and defending a competing positive account. Such behaviour suggests that undermining public confidence is the principal objective, not advancing our understanding. The fact that mainstream science is unable at present to convincingly explain or account for certain data is not in itself a reason to embrace, or even express sympathy for, particular alternative theories. The latter establish their own credentials by providing positive, reasonable arguments. No-one was convinced that atoms have nuclei, that gametes have half the number of chromosomes of somatic cells, that the continents drift relative to one another, that species evolve slowly over time and so on simply by advancing critical arguments against prevailing ideas and attitudes. As Miriam Solomon notes, genuine scientific dissent requires positive empirical evidence for an alternative theory. To suppose that the explanatory inadequacies of one theory are in themselves reasons for preferring a competing theory is to commit the fallacy we called false choice.

Engaging only public audiences

Our final indicator that a controversy is created rather than authentic is evidence that more effort is being directed towards convincing the public of some alternative perspective, rather than the relevant scientific community. If what matters to a group is *public* support for their cause, or *public* confusion with respect to some issue, then we would expect efforts to be directed towards experts only to the extent that it serves the primary objective. By definition, a created controversy surrounds an issue over which the majority of qualified experts are in broad agreement. Now consider a group that starts raising objections against certain scientific claims, but these objections are regarded by those qualified to judge as muddled or irrelevant. Rebuffed by the scientific communities, the group might, in some cases, persist with their conviction that something about mainstream scientific research is flawed. They might feel justified, even obligated, in their plight to educate the public. Alternatively, such groups might not care about the opinions of the scientists and will thus readily and more knowingly commence a program of deception and distortion. In either case, attention shifts towards influencing public attitudes.

Not only will efforts be skewed towards persuading the public, we can expect those efforts to be non-trivial. Most genuine scientific controversies aren't all that interesting to most people, and so, while they might receive some media attention, as long as the controversy remains unresolved there will be little reason to continue drawing the public's attention to the matter. However, when the experts are in agreement, then those conclusions can be expected to gradually disseminate more widely, at least in the absence of an ongoing campaign that's marshalled in opposition. Maintaining the appearance of uncertainty requires vigilance. Thus, the more attention an apparently scientific controversy receives within the public domain the more suspicious we should become that it's only receiving so much attention because there are groups committed to keeping the appearance of controversy alive.

The three indicators I've described are suggestive, not definitive. Those independently motivated to discredit scientific ideas might be capable of unearthing valuable evidence that advances their cause. Criticizing existing ideas is central to scientific inquiry, so drawing attention to potential deficiencies is not unimportant, and perhaps some genuine scientific controversies would receive ongoing media attention, particularly if they have social significance. Nevertheless, attending to the possibility of self-serving motives, the role of uncertainty and the spheres in which debates take place can each be instructive. Finally, beyond these three indicators – and as we'll see in subsequent chapters – sometimes we have access to more direct evidence that a controversy is created; that is, strong evidence that the overwhelming majority of qualified scientists are in agreement over issues where public opinion is deeply divided. On many issues, national science academies, scientific organizations, institutes and societies release public statements designed to publicize the considered opinion of that group's members. Clearly we should pay close attention to such pronouncements.

How to respond

We have spent time discussing what manufactured controversies are, how they can be created and how we might hope to spot them. Thus far, however, we have said nothing about how we should respond once we're convinced that a given controversy is contrived. If we were simply to recommend those conclusions that are agreed upon by the majority of experts without qualification, we might give rise to the unfortunate implication that the sciences are decided by consensus. Those who object to particular scientific theories or conclusions on ideological grounds will happily point out that by sciences' own standards conclusions are supposed to be established through argument and evidence, not by popular vote. However, there is a difference between trained, qualified experts accepting the conclusions, methods and assumptions of their discipline because the majority has already declared their allegiance to that view, as opposed to novices accepting theories because the majority of scientists are in broad agreement. The former is far more problematic than the latter.

In Chapter 4, we considered the idea that epistemic communities are strengthened insofar as they reward those members who correct mistakes within that discipline, or argue plausibly for novel ideas. Maintaining a critical, demanding attitude is important for scientists. However, this doesn't entail that people outside the community can't reasonably defer to the consensus opinion in many circumstances. The advice here is only that, without very good reason to the contrary, we should place some degree of confidence in the proclamations of those who are better positioned to reach reliable conclusions, particularly on issues that have received enormous attention and particularly when something approaching a consensus among the relevant community exists. The attitude is a familiar one. We trust experts all the time. We generally suppose that doctors know more about our ailments and injuries, that lawyers know more about the law, mechanics know more about engines, architects know more about building safety standards and regulations and so on.[7] Assuming that the majority view among working scientists is the most sensible view for the layperson to adopt is advice that shouldn't seem like a very radical suggestion. However, if we don't like that suggestion, we can instead suppose that scientific conclusions should not be accepted simply because they are endorsed by the majority of experts (even in the case of the novice); nevertheless, adopting the majority attitude is sensible *in the absence of extraordinary reasons to suppose the majority is wrong*. In other words, if we are going to reject mainstream scientific conclusions, we should have extremely good reasons for doing so. Articulating what constitutes *an extremely good reason* is challenging, but we make inroads by recognizing some reasons that patently don't qualify. Here are a few *poor* reasons for rejecting scientific consensus.

- The fact that many others harbour doubts about the expert opinion isn't a good reason for supposing the doubts are well grounded, because the popularity of a belief is not a reliable guide to its reasonableness. Members of past civilizations insisted that Earth was both flat and at the centre of the universe. We should also remember the bandwagon effect: opinions seem to us more reasonable if they are repeated often enough.
- The fact that some appropriately qualified scientists have doubts isn't a good reason for supposing their doubts are well grounded. Dissenting voices are heard in all scientific disciplines. Some people like to be contrary or become wedded to ideas that have been discredited. Even scientists can be obstinate, unable to appreciate the import of an argument that is transparent to (almost) everyone else. If a very strong majority of qualified experts find the evidence for some conclusion convincing, then generally the attitudes of the minority should be accorded weight proportional to their numbers.
- The fact that we can't personally, and *impromptu*, answer an objection that someone raises against the prevailing scientific view isn't a good reason for supposing that good answers aren't available. Similarly, the fact that we have unanswered questions or doubts about a particular scientific issue is a poor reason for supposing that very good answers to these questions aren't available. Our personal incredulity cannot be a good reason for rejecting a given conclusion, scientific or otherwise. If the many hundreds and thousands of qualified individuals find the evidence and arguments convincing, then it takes an extraordinary degree of hubris for the novice to dismiss the entire edifice on the basis of uninformed doubts.
- The presence of genuine anomalies, surprises and outliers is not always a good reason for heightened scepticism. These are part of science, tolerated because they are heavily outweighed by positive evidence and arguments.

[7] The creation of controversies might even be regarded largely as exploiting our tendency to trust experts. When celebrities, politicians and seemingly learned individuals speak in lecture halls, on radio or television, it is hard for us not to suppose that they are in possession of important information, even if their claims do conflict heavily with the scientific consensus.

- It is no justification of sceptical attitudes to appeal to our right to hold whatever personal opinions we prefer.

Having gathered good evidence that a controversy is created, thus does not properly reflect the opinion of the scientific community, and in the continued absence of any good reason for resisting the science, a sceptical attitude towards the science begins to look obtuse. Certainly we should admit that science is fallible, that the general theory might require revision, correction and perhaps partial retraction. But the mere *possibility* of systematic and fundamental error alone is not a good reason for supposing that it exists. We can't reasonably reject the consensus opinion of a community of experts on the basis merely that they might all be wrong.

Conclusions

If a slew of objections are directed toward particular scientific theories or conclusions, then it is hardly surprising that people will develop doubts and perhaps even acute reservations about them. There is a tendency to suppose that the sciences deliver facts and that facts aren't disputable. This tendency leads us astray. The appearance of scientific controversy might reflect nothing more than the sustained effort of groups who are desperate that certain ideas remain shrouded in uncertainty for as long as possible. The appearance of controversy is consistent with an overwhelming majority of qualified individuals having reached a broad consensus on the issue, due to a tremendous amount of highly relevant and reliable evidence that confirms those attitudes very clearly. Ignoring the best available information is risky, and thus mistaking well-established scientific conclusions for speculation or guesswork is something we must all strive to avoid. Speciously dismissing important results can affect individual behaviour, undermine support for policies that might have enormous social significance, and hinder scientific progress.

Created controversies are the result of employing certain strategies that exploit our vulnerabilities and are often motivated by financial or ideological factors. In previous chapters, we have attempted to reduce our vulnerabilities by reflecting on the nature of scientific inquiry, as well as work in cognitive psychology and critical thinking. In this chapter, we have drawn attention to the strategies that help create and keep controversy alive, as well as the role of problematic motives. The two most important consequences of a created controversy are practical decisions that are riskier and more dangerous than the alternatives and, second, a more profound level of scientific ignorance than we begin with. The latter highlights the fact that created controversies exploit our unfamiliarity with *particular* scientific methods, theories, evidence and results, as well as our misconceptions about how sciences work, but then inflates those confusions so that individuals start propounding conclusions that are violently at odds with well-established scientific opinions. Understandable personal uncertainty and

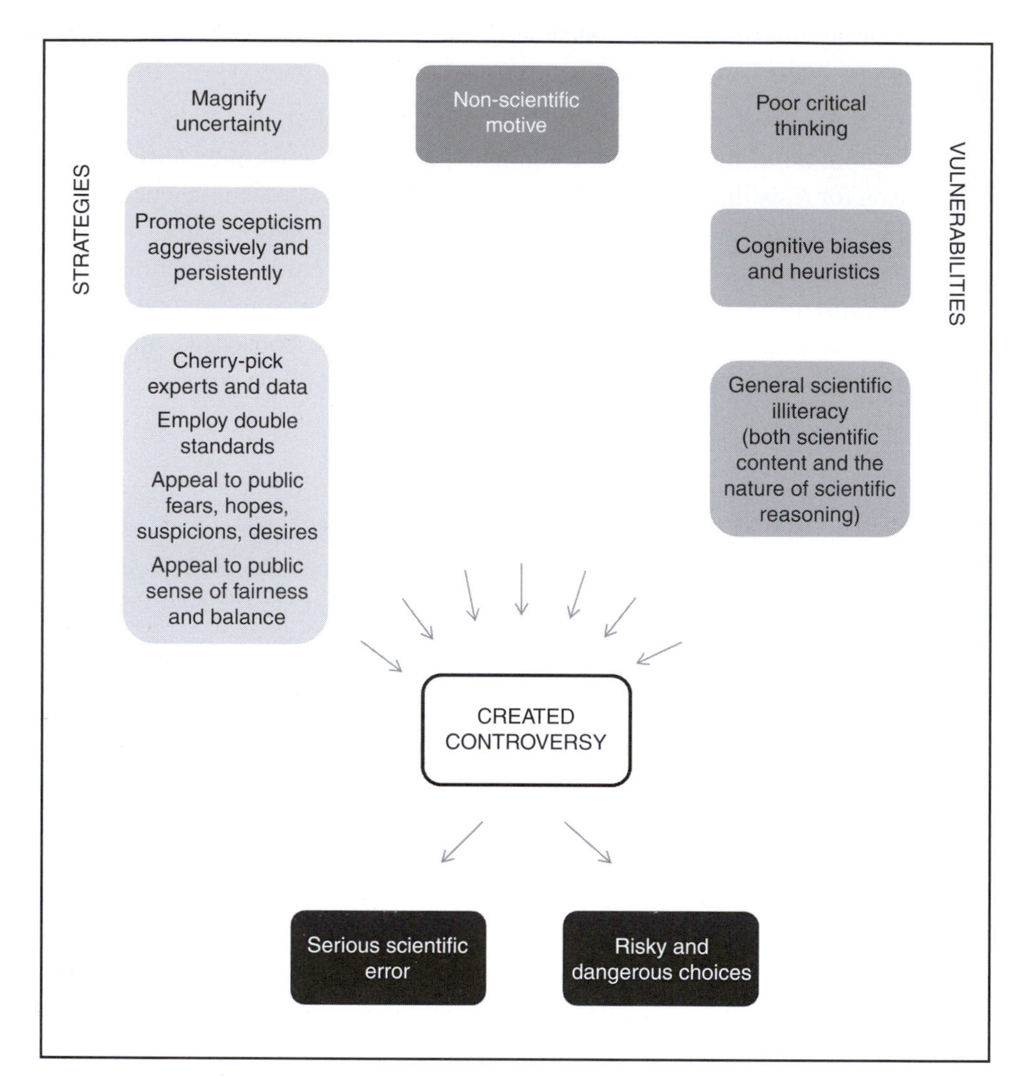

Figure 7.1 Causes and effects of creating controversies

ignorance surrounding the details of radiometric dating, how vaccines work, the difference between weather and climate, what distinguishes genetically modified food from conventional techniques of plant breeding and much more besides can each be cantilevered out until we start endorsing conclusions that are acutely implausible given available evidence. Knowing little about how the ratio of potassium to argon atoms within rocks can be measured and then utilized to provide reliable estimates for the age of the rocks is understandable, but when individuals dismiss centuries of study from geology, chemistry, physics, astronomy, biology and more and conclude that Earth is just a few thousand years old, something has gone profoundly and seriously wrong.

The discussion in this chapter has proceeded without attention to those issues that are often described as paradigmatic examples of created controversies. In the final section of the book, we will consider several such cases. A better sense for the details surrounding these topics should not only produce a better understanding of the general concept under consideration, but also a better grasp for affairs that hold enormous social significance.

Discussion questions

1. Why might it be better for groups to create doubt rather than simply *deny* scientific conclusions and evidence?
2. When considering created controversies, we can distinguish the empirical evidence that supports certain conclusions from the fact that there exists a scientific consensus surrounding the issue. Are these different types of evidence? Is one more important than the other? Is it fair for laypersons to rely on both? Are there other kinds of evidence that might also be relevant?
3. Why is it hard to distinguish genuine controversies from the appearance of controversy?
4. One version of the fallacy of appealing to ignorance can be understood as inferring from *we can't be certain that X* to *X is as likely to be false as it is to be true*. How important do you think this fallacy is, for purposes of understanding created controversies?
5. In Chapter 2, we sketched a more positive perspective on the problem of induction. How does that help us think about uncertainty within the context of created controversies?
6. Most people acknowledge the expertise of doctors, lawyers, engineers and so on. Are there reasons to treat communities of scientists any differently?
7. Are there circumstances in which we might identify a controversy as created, but where we might justifiably continue to deny the mainstream scientific view? If so, what are these circumstances?
8. If there remain unanswered questions within some field of inquiry, and a range of competing, somewhat plausible answers, should we conclude that all conclusions within that field are controversial? Is this an inferential leap that some people make?
9. Why is the cherry-picking of particular studies hard for the layperson to spot? What is the significance of this handicap? Can we prepare ourselves to notice cherry-picking?
10. Is complete consensus, on some scientific issue, an unrealistic standard? Why, or why not?

Suggested reading

Several books explore the distortion of scientific conclusions, principally from political, social and historical perspectives, for example Mooney (2006), Otto (2011) and Oreskes and Conway (2011).

Points to remember: Part II

1. While we're often not aware of it, our reasoning is predictably irrational:
   a. We preference arguments and evidence that conform with prior beliefs
   b. We are unduly influenced by what's easiest to recall
   c. What's easiest to recall is influenced, among other things, by what's repeated most often
   d. We place too much confidence in our own abilities and too much emphasis on the evidence we remember
   e. We are not good at admitting our vulnerabilities to cognitive biases
2. With any argument, we should reflect on the quality of the evidence being utilized, the extent to which it supports the conclusion and the possibility that competing explanations are better supported by available evidence (which may include a lot of evidence that we're either forgetting or are unaware of).
3. Reliable argument evaluation is extremely hard work. Only by carefully developing good critical thinking skills, learning to identify informal fallacies and becoming better informed can we hope to reach sensible judgements on complicated issues.
4. The appearance of uncertainty within some scientific community may not represent actual uncertainty.
5. If a scientific community is in broad agreement on some issue, then only truly extra-ordinary reasons should cause us to dissent. The more carefully a theory or conclusion has been reviewed, and the stronger the consensus, the more we should be wary of rejecting that theory or conclusion.
6. That anomalies, puzzles and lacunae are apparent is neither evidence that one theory is in crisis nor that an alternative is in ascendancy.
7. It's easy to cherry-pick data. The world is complicated, and we shouldn't be surprised that there are outliers. If a large number of studies are conducted and almost all gesture towards the same conclusion, then focusing attention on the outliers is deceitful.
8. It's easy to cherry-pick experts. People are complicated, including scientists, so we shouldn't be surprised that some defend conclusions that are regarded by the rest of the

scientific community as implausible and without merit. If a large proportion of scientists all agree on the basic issue, then focusing attention on the exceptions is unreasonable.

9. For ideological, political, moral or financial reasons some groups have motivation to *want* the science to be wrong, hence an incentive to look for ways in which it might be wrong and then publicize those findings. Such desires compromise objectivity.
10. Magnifying uncertainty is central to spreading public doubt. If we are suspicious that a group is only interested in inflating uncertainty, then we should be wary that they are hoping to create the appearance of controversy.
11. The fact that we don't personally understand every aspect of an issue is quite consistent with others having a very good understanding, and also with having very good reasons to support their explanations.

Part III

Exposing Created Controversies

8

Environmental scare: the case of anthropogenic climate change

There are many who regard anthropogenic, global climate change as the most urgent problem facing humankind. There are also critics who dismiss the whole idea as a hoax, who remain deeply sceptical of even the basic scientific conclusions and who therefore feel no motivation to even consider policies that would reduce carbon emissions. Polls indicate that public attitudes are decidedly mixed. A 2010 study revealed that 61 per cent of Americans believed that the planet is getting warmer, but 45 per cent agreed that '[t]here is a lot of disagreement among scientists about whether or not global warming is happening'.[1] With almost half the population supposing that there exists a genuine scientific controversy, it is worth exploring the accuracy of that judgement. Is the appearance of controversy a reflection of genuine scientific disagreement, or is this appearance better explained in terms of public ignorance of the relevant evidence, cognitive biases, poor argument evaluation and the agenda of groups that feel threatened by a scientific consensus? The chapter is not designed as an authoritative overview of the state of the science nor as an attempt to answer the many particular criticisms that have been levelled against the sciences surrounding climate change. We might nevertheless each hope to become better informed on the issue. Our framing concept of a created controversy will also prove instructive.

The first two sections of the chapter will survey the history of our understanding of the climate, then offer a greatly condensed argument for why scientists accept that human activity is responsible for recent unusual climate change. Providing even a brief overview will help us recognize that many of the objections raised against climate change are simply beside the point. For example, once we grasp even a shred of the science we'll realize that it is entirely irrelevant, for purposes of understanding anthropogenic climate change, that carbon dioxide is a *natural* gas, just as it's irrelevant that CO_2 comprises only a tiny fraction of Earth's atmosphere. Unfortunately, despite their utter irrelevance, prominent individuals have offered both these observations as apparent challenges to climate science.

[1] Leiserowitz et al. (2011).

In the third section, we'll consider influences on public attitudes towards climate science, including what appear the most common *given* reasons for remaining agnostic. Finally, we'll utilize the concept of a created controversy to test the extent to which the apparent controversy surrounding climate change should be regarded as illusory, rather than a reflection of genuine discord within the relevant scientific community. By the end of the chapter, readers should know more about the basic scientific evidence and arguments, hence will be better situated to evaluate certain objections and to understand both why scientists are convinced that anthropogenic climate change is happening and why it threatens to profoundly affect our lives.

A history of understanding the climate

The suggestion that the atmosphere has an influence on global temperatures, and thereby global climate patterns, is an old one. In the early nineteenth century, Joseph Fourier, a French physicist and mathematician, calculated on the basis of incoming and outgoing radiation that our planet should be considerably cooler than it actually is.[2] He suggested, in response to this discrepancy between prediction and observation, that the atmosphere might trap some energy and thereby make the planet warmer than it would otherwise be. The plausibility of the explanation depended, in part, on evidence that atmospheric gases have the capability to trap, or absorb, infrared radiation. Such evidence appeared towards the middle of the nineteenth century courtesy of physicist John Tyndall.

Tyndall was interested in ice ages. Observations gathered earlier in that century had convinced geologists that huge ice sheets had once covered vast areas of northern Europe and North America. What intrigued Tyndall was the possibility that he might identify those circumstances that could cause an ice age, or bring one to a close. Curious about a possible connection between global temperatures and atmospheric composition, Tyndall explored Fourier's suggestion that atmospheric gases might absorb infrared radiation. He confirmed that the majority of Earth's atmosphere cannot, but that certain trace gases do occur naturally in the atmosphere (mainly CO_2, but also methane, nitrous oxide and water vapour) and do absorb such radiation.

Modern science now fully endorses Fourier's conjecture: some naturally occurring atmospheric gases do absorb infrared radiation and thereby have a warming effect on our climate. These are known as greenhouse gases. What's known as the *natural* greenhouse effect is crucial for life as we know it. Without those gases, the climate would be considerably cooler. It is the *enhanced* greenhouse effect, due to anthropogenic greenhouse gas emissions, that is of concern to climate scientists and which we will return to later. The proportion of the atmosphere that absorbs infrared radiation is tiny. This led Tyndall to conclude that atmospheric changes were an

[2] The historical details in this section are taken from Weart (2008).

unlikely explanation for anything as significant as the rise and fall of the ice ages. Svante Arrhenius, however, a Swedish physicist, chemist and Nobel Prize winner, would later recognize something that Tyndall had not.

What Arrhenius realized, working towards the end of the nineteenth century, was that small changes within the atmosphere might have large effects due to *feedback* mechanisms within the climate system. For example, suppose atmospheric CO_2 increased slightly, perhaps through volcanic activity. Temperatures would thus increase slightly, and this would likely produce more evaporation off large bodies of water, which would introduce more water vapour into the atmosphere. Since water vapour is also a greenhouse gas, then greater water content in the atmosphere would have a further warming effect, hence more evaporation, more water vapour in the atmosphere, and so on, and so on. Small initial changes could potentially, through this and other positive feedback mechanisms, have much greater significance than first appearances might suggest. Feedback mechanisms continue to play a central role in scientists' understanding of climate change. The huge number of such mechanisms is one indicator of how complex the climate system is.

Arrhenius estimated how average global temperatures would change in response to changes in the quantity of atmospheric CO_2. The data available to Arrhenius was not particularly reliable, and the puzzle was soluble only by ignoring a host of complicating factors. Undeterred, Arrhenius considered the question worth pursuing and eventually concluded that halving the amount of CO_2 in the atmosphere would reduce average global temperatures by around 8°Fahrenheit. There were, however, several important reasons to doubt that Arrhenius's work could really explain anything about climate patterns. First, there was laboratory-based evidence which suggested that adding or removing atmospheric CO_2 wouldn't affect how much infrared radiation was absorbed, hence would have no effect on climate patterns. It was also recognized that the amount of CO_2 in the atmosphere was dwarfed by the quantities of CO_2 in minerals and oceans. Thus while it was acknowledged that burning coal put CO_2 into the atmosphere, it was quite conceivable that the excess wouldn't *remain* in the atmosphere for long. Perhaps the carbon would quickly migrate into the oceans. Finally, Arrhenius's work relating atmospheric conditions to climate patterns was just one of a variety of approaches to studies of the climate that were pursued during the first half of the twentieth century. Geologists described how mountain ranges are raised and subsequently eroded, which would influence wind and rain patterns, and thereby the climate. Other scientists pursued connections between the climate and ocean currents, solar activity, volcanic activity and changes in Earth's orbit about the Sun. The greenhouse effect came to be regarded as less important than several other factors, for purposes of understanding global climate patterns.

In the 1950s, two scientific papers appeared which suggested that the role of CO_2 should not be too quickly overlooked. First, Gilbert Plass argued that adding more CO_2 to the atmosphere could in fact increase average global temperatures, contrary to those experiments that had provided one key challenge to Arrhenius's work. Plass argued from the fact that the atmosphere is layered, and that different regions have different properties. Plass's observation was essentially that more CO_2 in the upper layers of the atmosphere would have a greater warming effect and that this would in turn affect temperatures at lower levels. At around the same time, Roger Revelle, an oceanographer, was considering questions about ocean chemistry and, in particular, the ocean's capacity for absorbing CO_2 from the atmosphere. Revelle concluded that the ocean would absorb far less atmospheric CO_2 than had been previously assumed.

Thus by the 1960s there was reason to suppose that adding more CO_2 into the atmosphere could have a warming effect, and there was at least no very compelling reason to deny that the emissions from fossil fuel burning would increase concentrations of atmospheric CO_2. Over the last five decades, there has been remarkable progress within the climate sciences, along a large number of dimensions. There have been important new sources of data, such as those gathered from ice cores, plant and pollen fossils, and archaeology, as well as vastly improved methods for measuring atmospheric composition, temperature, sea level, precipitation patterns, Arctic sea ice loss and glacial retreat. This has been aided by important technological advances such as satellite imagery.

There has been enormous growth in computing power, which has facilitated far more useful modelling techniques. Global climate models have been pivotal to the progress climate scientists have made. Developments in astronomy have led to important discoveries about the climate and atmosphere of other planets in our solar system. From physics and our understanding of fluid dynamics, to chemistry and the properties of atmospheric gases, to ecology and the effects of climate change on species and ecosystems, scientists from a wide variety of specializations now contribute to improving our understanding of our changing climate, its causes and the consequences of further change.

The history of climate science quickly dispels any notion that our environmental sciences arose in service to an environmentally motivated political movement. That CO_2 absorbs infrared radiation is beyond dispute and easily demonstrated with a bench-top experiment. The fact that CO_2 occurs *naturally* in the atmosphere is not a reason to doubt that too much of it can be harmful; there are many toxins and venoms that occur entirely naturally but are nonetheless fatal. The fact that CO_2 makes up less than 0.04 per cent of the atmosphere is a similarly poor reason to deny that it might have harmful effects. (Less than a gram of some venoms can be deadly to humans.)

Burning fossil fuels emits CO_2 and even by the 1960s there existed no obvious reason to deny that, other things being equal, adding more CO_2 to the atmosphere would increase average global temperatures. All these claims have been settled in broad outline since the mid-twentieth century or earlier. Accepting them does not entail anthropogenic climate change. Nothing that has been said in the chapter thus far rules out the possibility that, as a species, we simply don't produce enough greenhouse gases to have any discernible effect on climate patterns. Or it might be that CO_2 doesn't accumulate in the atmosphere, or perhaps the effects on climate patterns of increasing atmospheric greenhouse gases are so heavily dominated by variations in solar output, volcanic activity, ocean currents, orbital eccentricities and so on that there's no danger of human emissions ever having any effect. The plausibility of each of these scenarios requires a little more familiarity with the evidence in support of human-induced climate change, to which we now turn.

A short argument for anthropogenic climate change

As with other created controversies, what lingers around questions of climate change is a general cloud of uncertainty. Public scepticism is attributable, in part, to the impression that scientists themselves don't agree on the central issues. Accompanying this belief about the scientific community is a more general and ambiguous set of ideas that prevent any clear message concerning climate change from emerging. As individuals, we are unable to sort through and evaluate the many different ideas, so perhaps we find it comforting to suppose that no-one really knows whether human activity is influencing the climate. When attempting to evaluate issues of great complexity, we should be extremely wary of placing too much confidence in our personal ability both to appreciate the range of evidence and arguments that have been advanced within a scientific community and the problems that may have been unearthed with objections that we might find convincing.

We can't all become climate scientists, but it's helpful to sketch a basic argument for human-induced climate change. Not only will it help us recognize irrelevant objections and criticisms for the distractions that they are, it can also focus our attention in a way that clarifies the source of our own uncertainties, as well as those of friends, community members and public commentators. The basic argument that I'll outline for anthropogenic climate change and the importance of responding consists of four main claims:

1. The planet is undergoing unusual climate change.
2. Atmospheric greenhouse gases, and carbon dioxide in particular, are becoming more abundant due to human activity.
3. To the best of our knowledge, recent unusual climate change is in very large part attributable to the increases in atmospheric greenhouse gases. Given the nature of such gases, this shouldn't come as a big surprise.
4. The same methods that connect recent climate change with human activity can be utilized to provide projections about the next few decades of climate change. Those projections are of a climate that will create significant challenges for future generations, unless serious actions are taken to mitigate these effects.

Doubts can be, and have been, directed against each of these claims. We'll consider some of the specifics shortly. We'll start with some of the evidence that supports them.

Climate change

Most of us would look to the surface temperature record for the most important evidence of unusual climate change. It's not an ambiguous record. According to the U.S. National Climatic Data Center the fourteen years from 2001 to 2014 are all among the sixteen hottest years on record. Assessing global averages is not a trivial undertaking, and different data sets and analyses report slightly different results. Nevertheless, major studies that draw on enormous data sets from a variety of sources, paying close attention to discrepancies and possible errors, all converge on the same basic conclusion that average global temperatures are steadily increasing. The 1980s was the warmest decade since records began in the mid-nineteenth century. The 1990s was even warmer. The first decade of the new century was warmer still, and the average temperature from 2010 to 2014 even warmer again. Our efforts to measure temperatures reveal a clear pattern, but there are additional observations that provide important independent support for the claim.

Over the last few years, there has been an attempt to shift the discussion away from the concept of *global warming* and towards the concept of *climate change*. Climate scientists deem the latter term a more appropriate object of discussion for two reasons: first, the phrase *global warming* might imply that the main concern for future generations will be a slightly warmer climate, but this underestimates the range of challenges that will likely be faced, including rising sea levels, droughts, more and stronger storm systems and a decline in agricultural productivity. The second reason that the phrase *climate change* is more apt concerns the variety of evidence that illustrates recent climate change. For example, evidence has been gathered in recent decades of widespread glacial retreat over the last century (see Figure 8.1), rising sea levels going back even further, warming oceans, thawing permafrost and rapidly thinning summer ice in the arctic. Efforts to measure these data are sophisticated, comprehensive and improving. In some cases, such as sea level rise, the message from the data is clearer than that provided by the temperature record (see Figure 8.2). More indirect evidence of warming further strengthens the argument. Some plant and animal ranges are shifting poleward and into higher altitudes. Growing seasons are lengthening in northern latitudes, especially at higher altitudes. Plants flower earlier, migratory birds arrive earlier and insects emerge earlier. These phenomena are unsurprising if we admit to rising temperatures, but collectively would create a huge number of puzzles if we insist that the temperature record is mistaken and that temperatures are stable. Our planet is getting warmer.

The warming trend of the last few decades is clear, but there is also evidence that modern temperatures are the highest our planet has experienced for tens of thousands of years. (The planet has been warmer, if we look back far enough, but that's little

Figure 8.1 Two images of Pederson Glacier, Alaska, taken sometime between the mid-1920s and early 1940s (top) and August 2005 (bottom), illustrating significant retreat over the twentieth century. Similar pairs of pictures taken from a wide variety of geographical locations provide striking illustrations of one consequence of global warming. Photo credit: U.S. Geological Survey, Department of the Interior/USGS, U.S. Geological Survey/photo by Bruce F. Molina.

comfort if those temperatures would precipitate enormous costs to human life.) Furthermore, there is evidence that the current *rate* of change is also unusual. This is important because the quicker a climate changes the more difficult it is to adapt. Ambitious, large-scale, technically involved efforts to better describe how the climate has changed are being conducted every year, and they will continue. Some of these studies will lead to some revisions of prior conclusions. There is no good reason to suppose, however, that such studies will overturn the central observations.

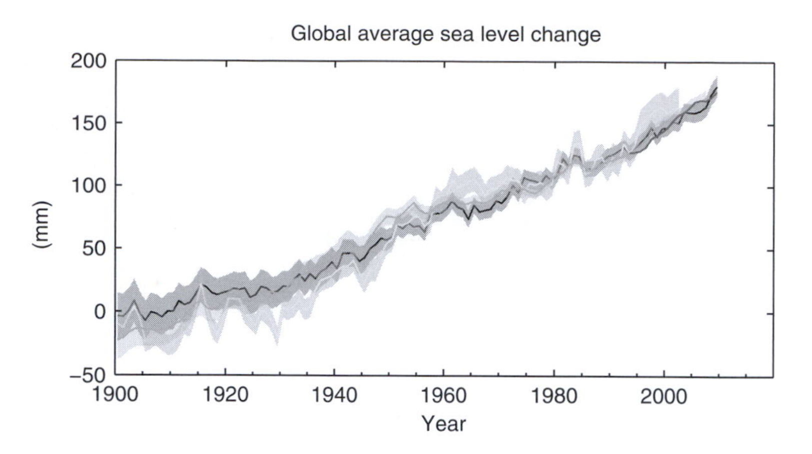

Figure 8.2 Combines three distinct data sets, each aimed at measuring global mean sea level rise. Average GMSL rise between 1901 and 2010 is estimated at between 1.5mm and 1.9mm per year. The rate from 1993 to 2010 is higher, 2.8 to 3.6 mm per year. Source: www.ipcc.ch/report/ar5/wg1/docs/WGI_AR5_2013_Poster.pdf.

Admittedly we do not see a surface temperature record of smooth continuous change, where every winter is a little milder than the one before, and every summer breaks the records that were set just twelve months earlier. However, this is entirely consistent with the predictions of climate scientists. No-one denies that many entirely natural phenomena influence the climate. A steady and uniform increase in CO_2 isn't expected to produce a steady and uniform increase in average temperatures. The thesis we're concerned with is that every year is warmer *than it would have been* were it not for the enhanced greenhouse effect. The reasons the vast majority of scientists accept that conclusion will be presented shortly, but it's important to be clear on what is being argued for.

The fact that these changes can't be directly experienced by us is also worth emphasis. We experience climate on an extremely local scale. Our memory is less reliable than written records. We improve our reasoning and arguments by appealing to comprehensive and rigorously analysed evidence, rather than anecdotal impressions. That certain *regions* experience milder summers, or more severe winters, doesn't itself contradict global warming. Of course, acknowledging that our planet is getting warmer, that sea levels are rising, glaciers retreating, rain patterns changing and so on is not an explanation for the trends. Before surveying explanations, however, let's turn to evidence that atmospheric greenhouse gases are also increasing.

Greenhouse gases

In the mid-1950s, a program was developed to monitor levels of atmospheric carbon dioxide. Stations were established in Hawaii and Antarctica, although the latter was closed in the 1970s due to budget cuts. Developed and overseen by Dave Keeling, the

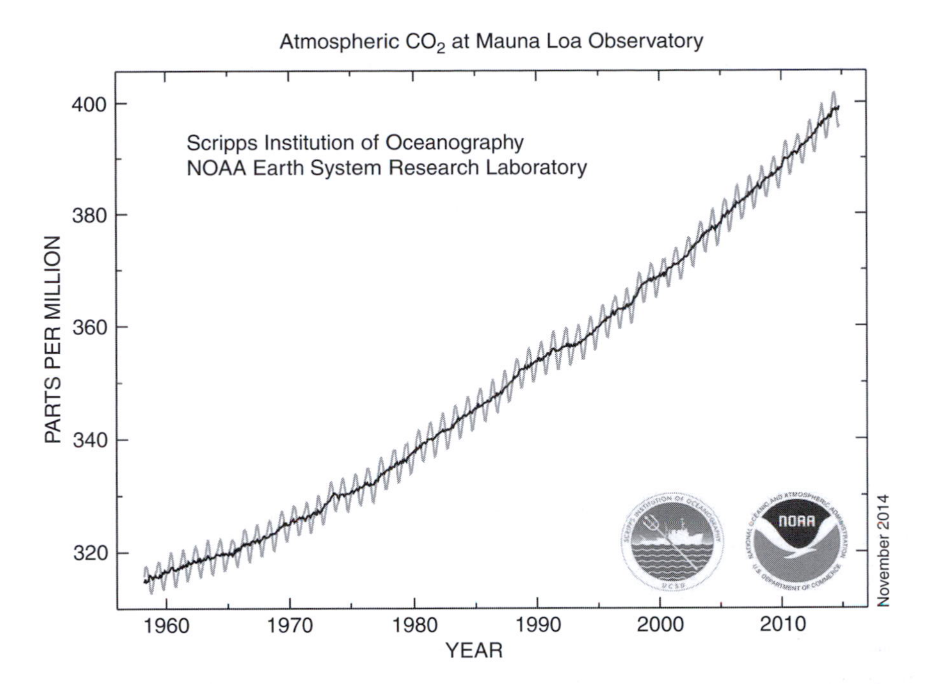

Figure 8.3 The Keeling curve, measuring atmospheric CO_2 since 1958. Source: NOAA ESRL Global Monitoring Division, Boulder, Colorado, USA

measurements that were gathered through this program have become emblematic of global warming and provide strong evidence of a robust trend (see Figure 8.3). When the initiative began, atmospheric levels of CO_2 measured around 315 parts per million (ppm). By 2013 these levels were very close to 400 ppm, a 25 per cent increase.

Another great advance in understanding our climate emerged with the extraction of enormous ice cores from ice sheets in Antarctica and Greenland. Ice cores provide a remarkable record of past climates. This is because in some regions summer temperatures never get above freezing, hence each year's fresh snowfall covers the previous year's. These layers build up, in some cases for thousands of years. Drilling down into an ice sheet, extracting a core and studying the contents of air bubbles trapped in the ice reveals information about both the composition of the atmosphere at the time the ice formed, and the temperature. Ice cores have enabled climate scientists to reconstruct records for the past half a million years. One of things we've learned is that atmospheric levels of CO_2 have not only been steadily rising for the last few decades, but they are higher now than at any other time in the last several hundred thousand years.

There are two major sources of anthropogenic greenhouses gases. The first, most obvious and most important is the burning of fossil fuels. Billions of tons of CO_2 are released annually from fossil fuel consumption.[3] The second major source occurs

[3] www.worldwatch.org/global-fossil-fuel-consumption-surges.

through changes in land use. Rainforests, for example, absorb large quantities of CO_2, so when these are cleared for urbanization or agriculture the change results in more atmospheric CO_2. Not all anthropogenic CO_2 ends up in the atmosphere, but the Keeling curve establishes that atmospheric levels are increasing. If there was any doubt about the origin of the excess, carbon dating provides the answer. There are three carbon isotopes that occur naturally in Earth's environment, ^{12}C, ^{13}C, and ^{14}C. Carbon-14 is radioactive, so it decays. It has a half-life of 5,730 years. Fossil fuels are sufficiently ancient that all except negligible traces of ^{14}C have long since disappeared. From these two claims it follows that the burning of fossil fuels should put a disproportionate amount of ^{12}C and ^{13}C into the atmosphere. In the 1950s, Hans Suess, a pioneer in the new field of carbon dating, predicted that over time we should observe the proportion of atmospheric carbon-14 declining. This decline, known as the Suess effect, is also well established.[4]

Carbon dioxide is not the only greenhouse gas that is becoming more abundant within the atmosphere. Methane is a naturally occurring greenhouse gas, but various human activities also produce it, and atmospheric levels of methane have increased significantly since the industrial revolution. The same is true of nitrous oxide. In addition, several synthetic gases present in the atmosphere are known to absorb infrared radiation. The climate is undergoing unusual changes, changes that are associated with warming, and atmospheric levels of greenhouse gases (those that have the capacity for absorbing heat) are increasing. Of course, we might still worry that the evidence falls short of establishing that the emissions are *causing* climate change. Evaluating this connection requires a different kind of argument, and it is the development of global climate models that have filled this need.

Global climate models

Certain kinds of models are very familiar. We've all admired model planes and trains, or miniature dioramas that depict a room, building or geological feature. We recognize these objects as models because they represent certain key features of the target objects that they're designed to represent. Models won't perfectly depict every aspect of the target object, but they can nevertheless help us better understand something about the structure, style or form of the object we're modelling. As well as objects, we can also model *processes*, or dynamical systems. Given the current state of a system, we can utilize mathematical equations to evaluate how that system will change over time. For example, Newton's equations of motion allow us to model the behaviour of an object that's free-falling towards Earth's surface. The equations enable us to evaluate the position of a moving object, at particular times, under specified forces.

[4] The Suess effect is also used to describe the depletion of atmospheric ^{13}C, as a result of fossil fuel burning. The explanation here concerns plant life's greater affinity for ^{12}C (Pilkey and Pilkey (2011, 5).

We can thereby represent, or model, the motions of objects. As with physical models, a dynamical model can further our understanding despite not accurately representing all aspects of the target.

Climate models are designed to help us understand patterns in temperature, precipitation, atmospheric pressure and so on over long periods of time. Climate sciences are thus importantly different from forecasting the weather. According to some commentators, the distinction between weather and climate is perhaps the most important distinction for purposes of relieving the uncertainties of a sceptical public. Weather forecasts are predictions for the temperatures, precipitation, winds and so on for a particular geographical region, at a given time of day. Climate projections concern long-term *averages* of these properties. It is extremely challenging to accurately forecast the weather even a week or two in advance, but this needn't impede our ambitions to successfully predict future climate change. We can be extremely confident that average temperatures in the northern hemisphere will be warmer next July than they'll be next December, for example. Predicting climate change is about predicting long-term trends, and this doesn't rely on an ability to make very accurate short-term forecasts.

Long-term climate patterns become better understood as we learn more about the climate system, that is, the many varied and subtle interactions between distinct aspects of the atmosphere, oceans, cryosphere, land and living organisms. Our planet is constantly being bombarded with solar radiation, some of which is reflected off the atmosphere and back into space, and some is absorbed by the atmosphere. Of that which reaches the Earth's surface, some is reflected back into space, and some is absorbed by land and oceans. As the land and oceans are heated, they emit infrared radiation. Over 99 per cent of the Earth's atmosphere is transparent to such radiation, but the greenhouse gases absorb that radiation and thereby have a warming effect on the Earth's temperature. Incoming solar radiation can vary, sometimes due to variable output from the Sun, but also because of variations in the Earth's orbit. The Earth's axis currently tilts at 23 degrees relative to the plane in which it orbits the Sun, but that angle changes over a period of thousands of years. Similarly, the shape of the Earth's orbit varies over lengthy periods, from an almost perfectly circular orbit to one that is slightly more elongated. These variations in the Earth's orbit are well correlated with ice ages.

Mountain ranges affect wind patterns, and thereby precipitation patterns. Warmer climates increase cloud coverage, which affects both the amount of solar radiation that is reflected back into space but also the amount of infrared radiation that is absorbed within the atmosphere. Different parts of the Earth's surface reflect or absorb different quantities of solar radiation. Ice and snow are highly reflective, but a warmer climate reduces total ice and snow cover, which means less solar radiation will be reflected back into space and more will be absorbed, so the planet gets warmer still. The relationship between temperature and ice cover is an important feedback mechanism.

The atmosphere and oceans are three-dimensional, where changes at their interface may be quite different from changes at the ocean depths and upper echelons of the atmosphere. The climate system is thus an enormously complicated web of interrelations and interactions. Changes to atmospheric composition, for example, affect other components of the climate, and changing these components will have further impacts including, perhaps, additional changes to the atmosphere. Our understanding of the many detailed interactions within the climate is forever improving and contributing towards a better grasp of global climate patterns and their causes.

The most pressing application for global climate models is to help us understand how temperatures and sea levels, for example, are likely to change, given certain projections about future carbon emissions. Climate models have become increasingly refined over the last few decades. The most comprehensive models being developed and utilized are three-dimensional general circulation models. The Earth's atmosphere is treated as a dynamic fluid. The models utilize mathematical equations that describe the physics of changes and processes in the atmosphere and oceans, as well as on land. These equations are solved for a three-dimensional grid over the planet. The amount of data, and complexity of the models, means that solving these equations requires an enormous amount of computing power. Climate science as it is currently conducted would not be possible were it not for the exponential growth in computing power that the computer sciences have achieved in recent decades.

The value of any dynamic model lies in its capacity to accurately represent how the target system evolves over time. The reliability of climate models can be tested against our knowledge of past climate conditions and changes. A volcanic eruption spews ash and carbon into the atmosphere in quantities that can be estimated and where the effects can be accurately measured. A climate model can be tested by introducing information about the quantities of carbon, sulphur and particulates released by the volcano, then observing what effects the model predicts. If those predictions align with our observations of what in fact took place, we have some reason to think that the models represent how the climate responds to at least some changes in at least some conditions. This method for testing models is known as hindcasting. The more of these tests a model passes, the more confidence is placed in those models. Their growing sophistication and the extent to which they accurately model past climate changes is evidence of both the credibility of modern climate sciences and the importance of greenhouse gases for purposes of understanding recent climate change.

When global climate models are informed of increasing greenhouse gas emissions, their predictions are far closer to observed climatic changes than when that same information is absent. Figure 8.4 illustrates global climate models' capacity to represent temperature anomalies when they are both informed and not informed of changes in atmospheric composition. Changes in sea levels are similarly successfully modelled on the assumption that atmospheric greenhouse gases are rising, but not

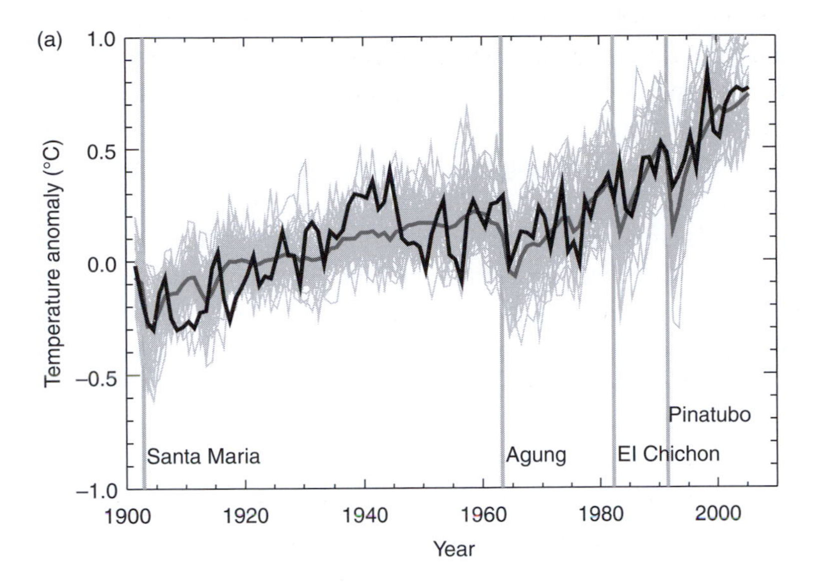

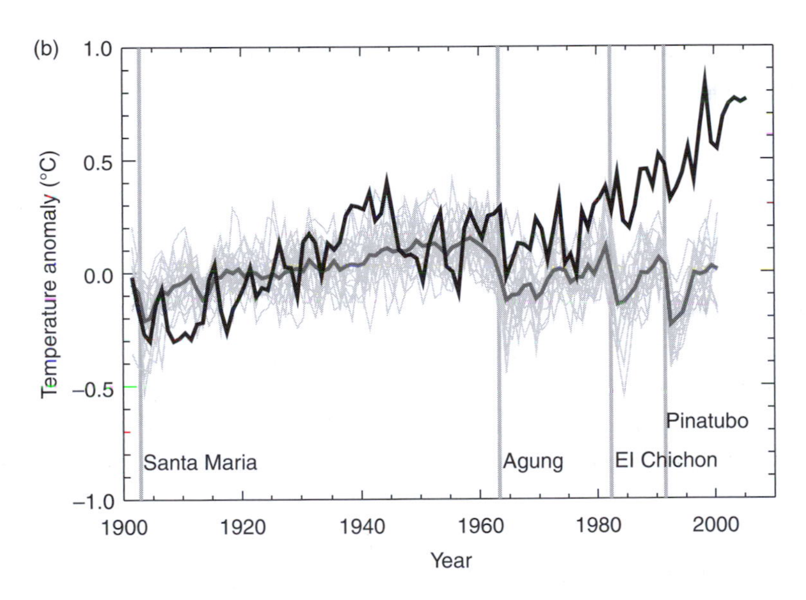

Figure 8.4 The thickest black line, which you'll notice is the same in each graph, represents global mean surface temperature anomalies, relative to the period 1901 to 1950. It provides a convenient means of seeing how average temperatures changed from 1901 to 2005. Model simulations of this same phenomenon from fourteen different models, and a total of fifty-eight simulations, are represented by the tangle of thin grey curves in the top graph. The pale tangle of curves in the bottom graph represents nineteen simulations from five models. The ensemble mean of both sets of simulations are also represented, as thicker grey lines that cut through the entanglements. The key difference between the two graphs is that the simulations represented in the bottom graph are based on natural variations only. The simulations represented in the top graph are based on both natural variations and atmospheric changes due to human activity. The four vertical lines denote four significant volcanic events; you can see (by attending to the thick black lines) that each is followed by a

otherwise. In other words, every known natural influence on climate patterns has been carefully analysed for purposes of understanding whether they can account for recent climate change, and none can. The conclusion is clear: recent climate change is explicable if, but only if, we recognize the role of changing atmospheric composition where, as we noted, the increasing levels of atmospheric CO_2 are the result of human activity.

Climate models have been in use for long enough now that we can compare earlier projections with subsequent observational data. Such comparisons are not entirely straightforward. First, projections must be based in part on estimates for future carbon emissions, so a projection will depart from observation if the quantities of emissions are under (or over) estimated. Second, because the climate is affected by entirely natural, but unpredictable phenomena, such as volcanic and solar activity, projections might depart from observations for reasons that again do nothing to discredit the accuracy of the model. There are inherent uncertainties within climate modelling. Nevertheless, recent reviews have concluded that the projections for global temperatures, offered by the Intergovernmental Panel on Climate Change (IPCC) in their 2001 and 2007 reports closely align with the actual temperature record. Furthermore, the inaccuracies within climate models need not portend good news. For example, over the last few years sea levels have been rising *faster* than IPCC projections.[5]

Future projections

Our models help us understand past climate change, but also help us better anticipate future changes. Global climate models don't predict a uniform increase in temperature for all regions. Some places will experience greater increases in temperatures than others, some places will get wetter, others drier. Island nations and coastal communities will become uninhabitable as sea levels rise, thereby generating refugees in numbers we have never before experienced. A more variable climate reduces agricultural yield, heightening concerns about how we will feed a rapidly expanding

Caption for Figure 8.4 (cont.)

noticeable, relative cooling. Both ensemble means capture this relation. More important, however, is the clear difference in fit between models that include information about human greenhouse gas emissions versus those that don't. In short, our understanding of recent temperature changes is good, but only if we don't ignore carbon emissions. Source: Fourth Assessment Report of the Intergovernmental Panel on Climate Change, S. Solomon, D. Qin, M. Manning, Z. Chen, M. Marquis, K.B. Averyt, M. Tignor and H.L. Miller (eds.), Cambridge University Press, figure 9.5, p. 684

[5] See Rahmstorf et al. (2012) for comparisons between IPCC projections and subsequent temperature and sea level records.

human population. Access to fresh water is a major concern, as is biodiversity. There is evidence that rapidly changing climates, such as ours, include more extreme weather – more violent storms, more heat waves, more droughts, floods and so on. These are not environmental concerns. These are concerns about human survival, human prosperity and the need to take actions that will help avoid colossal human tragedy.

The planet is getting warmer, and atmospheric greenhouse gases are becoming more abundant. Greenhouse gases absorb the energy that is emitted by the land and oceans, so it shouldn't be a surprise that their increasing presence within the atmosphere would have a warming effect. What we've learned from global climate models and our success in projecting climate patterns add enormous credence to this conclusion. We are left with little doubt that the climate will continue to change and in ways that will become increasingly dangerous. How we should respond to such threats gives rises to many difficult questions, but there is no reason to doubt the current state of the sciences. Critics will continue to dismiss or minimize aspects of the basic narrative, but we should be cautious that their arguments don't rely on cherry-picking data and experts, distorting the evidence, repeating objections that have already been answered, appealing to uncertainty, dismissing scientific conclusions as ideology or ensuring that the scepticism gains credibility through mere repetition.

The argument we have reviewed for anthropogenic climate change describes no more than a wisp of the total evidence and argumentation that have been assembled. Anthropogenic climate change is an extraordinarily well-researched issue with tens of thousands of research papers having now been reviewed and published. It would be extraordinarily naïve for laypeople to suppose they have any sensible appreciation for the depths that scientists have plumbed and the range of evidence that has been brought to bear. Better familiarity with the science than I have provided would no doubt be useful, but we should also take time to consider the sources of public scepticism surrounding the subject.

Reasons for scepticism

The most popular sceptical response to global warming supposes that recent climate change is due to natural variation. We're intimately familiar with the vagaries of the weather and thus seemingly feel comfortable extrapolating to long-term climate patterns. This attitude is probably enforced by our knowledge of past climates. We know something of ice ages and warmer periods, both in more recent and deep geological time. Evidence for natural variation is abundant, thus it might seem plausible to attribute unusual recent climate patterns to the same.

There are at least two significant problems with the proposal. First, appealing to natural variation is barely an explanation at all if it expresses only the idea that human behaviour is not to blame for climate change. As we've seen in this chapter, the climate is sensitive to a variety of changes, including solar activity, variations in the Earth's orbit, and volcanic activity. If the recent warming trend is explicable in terms of such natural processes, then it is entirely reasonable to ask *which one?* If no answer is forthcoming, we should worry that natural variability isn't an *explanation* for recent climate change but simply a refusal to countenance the possibility that human activity is to blame. Second, dismissing global climate change as part of natural variation is in considerable tension with some quite well-established facts: atmospheric greenhouse gases do, all else being equal, have a warming effect – this has been known since the mid-nineteenth century; second, the levels of atmospheric greenhouse gases are increasing; therefore, we appear to have reasons for expecting a rise in average global temperatures. Those who would describe recent changes in terms of natural variability owe us an explanation for why greenhouse gas emissions aren't doing what we have good reason for supposing they should do. Certainly, there might be answers to that question, but until convincing answers are forthcoming sceptics who perch on the reed of natural variability face a critical explanatory burden.

Appeals to natural variability are not the only objections that have been advanced against anthropogenic global warming. We'll consider some additional reasons below, but it is well beyond the scope of this chapter to answer all objections, or even to answer any particular objection in depth. By failing to seriously engage those arguments, it might seem that dismissing them is question begging and irresponsible. First, however, on issues like climate change it is impossible to anticipate and then satisfy the demands of almost any audience. Those intrigued by any of the particular objections to anthropogenic global warming can and should pursue them in the peer-reviewed literature or in the readings suggested at the end of this chapter. Far more importantly, however, conceding that there have been objections raised against climate sciences, which we are ignoring or paying at most scant attention, still permits two very different inferences.

If the overwhelming majority of qualified scientists are in broad agreement about what conclusions are supported by available evidence, then, with regard to particular criticisms and objections, either such scientists are irrationally and dogmatically ignoring problems with their conclusions, or they are fully aware of these objections but justifiably regard them as confused, mistaken or insufficiently probative to overturn what the majority of evidence recommends. The layperson should demand exceptional reasons to suppose the former inference is sensible. This deserves repeating and serious reflection: the layperson should demand exceptional reasons to suppose that specific anomalies, uncertainties and lacunae are evidence that scientific communities are irrationally and dogmatically ignoring problems with their theories and models. Our own idiosyncratic confusions surrounding the sciences of climate change are insufficient to suppose that the sciences are fundamentally flawed, as is the existence of others' objections that we're unsure how to answer. Criticisms have been raised against every single scientific idea ever articulated. We don't generally require answers to *every* objection before accepting what the experts announce. What does become pivotal, however, is whether there exists a consensus among climate scientists. Almost half the American public supposes that there is significant disagreement within the scientific community, but is that judgement accurate?

The Oregon Petition was first circulated in 1998. It claimed to have gathered the signatures of 17,000 scientists who agreed: 'There is no convincing scientific evidence that human release of carbon dioxide, methane, or other greenhouse gases is causing or will, in the foreseeable future, cause catastrophic heating of the Earth's atmosphere and disruption of the Earth's climate.'[6] A second version was circulated in 2007 and claimed to gather 31,000 signatures. The petition has attracted significant attention. It has also been widely criticized, with little attempt made to answer the objections. For example, according to the project website, individuals are approved to sign the petition 'if they have obtained formal educational degrees at the level of Bachelor of Science or higher in appropriate scientific fields'.[7] Scientific fields that are listed as appropriate include medicine, engineering, computer and mathematical sciences, biology and agriculture. Undoubtedly, there are individuals in all these disciplines that have expertise relevant to understanding the climate system and responding to the challenges it creates. However, the overwhelming majority of college graduates in these disciplines will never even take a course on climate science, much less conduct research or publish papers within climate science. The intention here is of course not to denigrate the academic qualifications of signatories, but to highlight the insignificance of the petition once we understand just a little more about its methods than attention-grabbing headlines may suggest.

The news that over 31,000 scientists deny global warming seems thoroughly noteworthy. Learning that 31,000 individuals deny global warming, but with no reason to suppose that they've even taken a college course on climatology, has a lot less punch. The petition project does claim that over 9,000 of the signatories have a doctoral degree, but the information that's publicly available makes it impossible to establish whether any of this group has significant, if any, training in climatology. This brings us to a second major problem with the project: independent efforts to verify that the names on the list are not forgeries have repeatedly failed. There is a reason that scientific communities insist on work being peer reviewed. We can place more confidence in work that has been checked by qualified editors and reviewers. The Oregon Petition doesn't meet the standards of credible scholarly work. As evidence that a significant number of appropriately qualified individuals deny the core components of anthropogenic climate change, the Oregon Petition is worthless. The question of whether the majority of scientists are in agreement is, nevertheless, an important

[6] http://ossfoundation.us/projects/environment/global-warming/myths/31000-scientists-say-no-convincing-evidence.

[7] www.petitionproject.org/qualifications_of_signers.php.

one. Several studies and surveys have been directed towards evaluating the issue. The results are unequivocal.

Naomi Oreskes thought to test whether there was a scientific consensus by analysing the abstracts of over 900 articles that had been published in refereed scientific journals.[8] Her intent was to identify the proportion of papers that either endorsed or rejected human-induced climate change. Oreskes identified zero papers that argued in opposition. Several subsequent studies have reaffirmed these findings. One follow-up study adopted a strategy very like that of Oreskes, but reviewed a whopping 12,000 papers. They found that 'Among abstracts expressing a position on [anthropogenic global warming], 97.1% endorsed the consensus position that humans are causing global warming.'[9] William Anderegg and some collaborators sought to investigate not just the degree of support among climate scientists, but also the degree of expertise of those included in their study. They concluded that '97 – 98% of the climate researchers most actively publishing in the field support the tenets of [anthropogenic climate change] outlined by the Intergovernmental Panel on Climate Change'.[10] Dozens of national science academies and major scientific organizations have endorsed the view that 'most of the global warming in recent decades can be attributed to human activities'.[11] The evidence for a scientific consensus concerning climate change is compelling.

The evidence is convincing, but it is important to remember why we care about the presence or absence of agreement. In particular, the fact that the overwhelming majority of qualified experts accept the core ideas associated with global climate change needn't function as the sole *reason* for supposing that those ideas are highly likely. For scientific communities, these ideas are regarded as true because the evidence compels that judgement. Those of us who have time and resources only to familiarize ourselves with a very cursory understanding of the central claims, and the evidence that supports them, may nevertheless be persuaded by that which we can assimilate. If we discover that the objections raised against those claims are unpersuasive, then this should enhance our confidence. Finally, for the layperson, the fact that the experts agree should function as a further, important type of evidence for us. Unless we can sensibly regard ourselves as experts in a given field, it is almost certainly not reasonable to dismiss the conclusions shared by the overwhelming majority of experts. It is nevertheless instructive to reflect on some of the styles of objection to climate change that continue to propagate.

Some sceptical responses towards anthropogenic climate change fail for reasons that basic critical thinking will reveal. People appeal to anecdotal evidence, or evidence that's irrelevant, or the opinions of those very few individuals who appear appropriately qualified but dissent from the consensus view. Such arguments will

[8] Oreskes (2004). [9] Cook et al. (2013). [10] Anderegg et al. (2010).
[11] www.skepticalscience.com/global-warming-scientific-consensus-advanced.htm.

convince no-one who is capable of evaluating them reasonably. A second important style of criticism appeals to uncertainties within the science. With a system as complicated as the climate, we shouldn't be surprised that these exist. Climate patterns are influenced by volcanic activity, which is very hard to predict. Future projections must also estimate future human carbon emissions, thereby introducing another obvious source of uncertainty. However, uncertainties surrounding some questions are fully consistent with enormous understanding on closely related issues. When critics appeal to uncertainty to undermine confidence in the entire state of the science they hope we will forget that truism.

A final variety of objection is answerable only through further engagement with the scientific literature. For example, it has been argued that recent warming is a result of increased solar activity, not greenhouse gas emissions. Second, data from ice cores have revealed that atmospheric CO_2 has become more abundant only several hundred years after ice ages have started to retreat, which might seem at odds with theories that describe atmospheric CO_2 as a cause of temperature increase. A particularly persistent thread in climate change scepticism is the notion that the planet hasn't warmed since 1998. What these objections share in common is that only a grasp of the relevant scientific data or theory suffices to answer the objections, but no layperson will be familiar with all the relevant information.

In fact, the peer-reviewed literature quickly dispels with all the objections, and many more. The last few decades have actually seen a decrease in solar output, while average global temperatures have continued to climb. For several decades, climate scientists have explained the cycle of ice ages primarily in terms of variations in the Earth's orbit. Changes in atmospheric CO_2 have never been regarded as the driving force behind the cycle of ice ages, but only an important feedback mechanism.[12] The cycle of ice ages isn't driven by atmospheric changes, but it wouldn't be as dramatic were it not for those changes. Thus, the fact the changes in atmospheric CO_2 aren't the primary cause of ice ages does not challenge the idea that increased levels of CO_2 are an important cause of global warming. Third, although 1998 remains one of the hottest years on record, a number of enormous statistical studies, based on multiple comprehensive data sets, have concluded that average surface temperatures have increased.[13] Additional objections have been raised, but it is unrealistic to suppose that we can all become sufficiently educated in the relevant sciences to answer all these objections for ourselves. It thus becomes important to recognize that our personal failure to answer a particular objection is a poor indicator that no answer is available. Perhaps the objections we're presented with have been answered by climate scientists, but those answers are dismissed, without good reason, or simply ignored by

[12] One suggestion is that as the oceans become warmer atmospheric levels of CO_2 increase, which then increases global temperatures yet further. If the ocean begins to cool, the reverse process occurs.

[13] Such objections also focus too narrowly on surface temperatures and thus ignore evidence of warming oceans. For more information see www.skepticalscience.com/global-warming-stopped-in-1998-intermediate.htm.

those who continue to deny climate change. Is it more reasonable to suppose that 97 per cent of qualified experts are all making the same mistake or to suppose that an answer to this question is available and we're just not aware of it?

Perhaps we think there's an enormous conspiracy among climate scientists, but this is hard to align with the facts we've already rehearsed. Carbon dioxide is a greenhouse gas, it is a by-product of burning fossil fuels, and it is becoming more abundant within the atmosphere. Average surface temperatures are increasing, sea levels are rising, and ice and snow cover is diminishing. Numerous climate models each suggest that recent climate change is attributable to increasing levels of greenhouse gases. If the community of climate scientists is involved in a huge deception, then what exactly are they deceiving us about? The problems in answering this question are in addition to the standard problems involved with any conspiracy theory: what's the motive for the deception, and how are so many thousands of people so effectively prevented from blowing the whistle? There are no good reasons to deny the consensus view, yet on radio, television, social media and the internet it is easy to discover the appearance of controversy. With arguments from sceptics being so readily available, we should worry that availability, overconfidence and confirmation biases will significantly impact public discourse. Some members of society have been denying climate change for so long that it will be extremely hard to replace their false beliefs.

A created controversy?

The significant work that's been done to establish the high degree of consensus, combined with public polls on the issue of climate change are sufficient for us to label this as a created controversy. The relevant community of experts recognizes available evidence as providing a very compelling argument for anthropogenic climate change. That consensus hasn't reached the public sphere, however. We hardly need review our three indicators of a created controversy. Nevertheless, it's hard to unlearn ideas that have been widely perpetrated, which continue to be repeated and which perhaps we've personally reiterated to friends, colleagues and family. It's hard for us to admit (even to ourselves) that we were wrong. It's even harder when there's little incentive to do so because the mistake is so commonly made. Thus, reflecting on the motivation of those who have done most to create the appearance of controversy, the amount of effort directed into keeping the controversy alive and the extent to which sceptics simply raise doubts (rather than offer alternative explanations) might all help us dislodge ideas that are just not sensible.

The deleterious motive for creating controversy is the financial concern of huge oil companies. Their annual revenues are in the hundreds of billions of dollars. According to a 2012 *Forbes* report, five of the world's seven most profitable companies are in oil and gas.[14] Several journalists have demonstrated that many of these companies

[14] http://www.forbes.com/lists/2012/18/global2000_2011.html.

regularly contribute significant funds to think tanks which challenge anthropogenic climate change and thereby keep the appearance of controversy alive. Complicating the issue is that some companies play both sides of the debate. One recent study observed that 'many companies [are] . . . supporting climate science and policy in some venues and opposing them in others'.[15] For example, energy and utilities companies continue to support various think tanks that spread highly misleading information about the state of climate sciences. These companies fund groups that undermine climate science while describing themselves publicly as both in agreement with climate scientists and committed to reducing carbon emissions and pursuing cleaner energy alternatives.

The idea that the science is controverted continues to receive attention, but the evidence for divided experts, problems with the data, problems with the models and so on doesn't change. These arguments have been answered, but the answers are ignored. Why are weak objections being repeated? Because there exists an ongoing campaign to mislead the public about the state of the science. We've observed that attributing recent climate change simply to *natural variations* is not a good explanation. Occasionally, reports will surface which attempt to relate global warming to particular alternative variables, such as solar output or CFCs, but these have failed to persuade those qualified to assess them. The majority of climate science scepticism is directed simply towards undermining the prevailing view. All three of our indicators point towards climate change as a created controversy.

Conclusions

The discovery of anthropogenic climate change was not the first time that scientists learned that industrial practices were causing significant environmental harm, and not the first time that the affected industries fought to undermine and suppress scientific conclusions while lobbying politicians aggressively for favourable treatment. In the 1970s and 1980s, there emerged compelling evidence both for the existence of acid rain and, quite unrelated, a hole in the ozone layer. In each case, the cause of these problems was clear, and the consequences of ignoring scientists were grave. In their important book *Merchants of Doubt*, Naomi Oreskes and Eric Conway demonstrate that a small number of ideologically motivated scientists successfully obfuscated the issues for purposes of manipulating public policy to the benefit of those industries whose profits were threatened.[16]

These individuals did not have training or qualifications in the relevant fields, and weren't actively researching the issues they sought to influence, but they had important political connections, ideas about the proper role of government and could rely on enormous financial support from the industries that were most immediately threatened

[15] Union of Concerned Scientists (2012). [16] Orsekes and Conway (2011).

by the scientific discoveries. They challenged the science in public venues, but the challenges were without merit. They ensured that uncertainty was widely propagated. Those same scientists are also linked with efforts to challenge the science surrounding climate change and, earlier, the tobacco wars. Oreskes and Conway provide an enormously important account of how industries have kept controversy alive within the public domain on issues that were scientifically settled.[17]

Jeffrey Sachs is the director of the Earth Institute at Columbia University. Sachs observes: 'The ultimate solutions to climate change are workable, cost-effective technologies which permit society to improve living standards while limiting and adapting to changes in the climate. Yet scientific, engineering, and organizational solutions are not enough. Societies must be motivated and empowered to adopt the needed changes.'[18] How societies and individuals can be effectively motivated to make changes is itself a difficult issue. People tend to be less motivated by future threats and risks than those that are more immediately apparent, a habit that is plausibly related to the availability heuristic we discussed in Chapter 5. Surveys conducted over the last few years have repeatedly demonstrated that Americans regard the problem of global warming as a very low priority for the country.[19] Individual and group failure to implement changes occurs because we assume we will have more time and more resources in the future, so we postpone adopting changes.

Not only is it hard to motivate people to act now, it's also hard to motivate them to do more than one thing at a time. Efforts to understand our perceptions of risk, and how uncertainty shapes our responses to issues like climate change, uncovered what has become known as the single action bias. Even if we do feel motivated to change, we have a tendency to perform just one action, such as insulating the attic or replacing a few incandescent light bulbs with the more energy-efficient compact fluorescent bulbs. Perhaps feeling that our single action means we've done our part towards helping the environment, we take no further actions. Thus, we see that there are important questions about how to motivate individuals, but a crucial step nevertheless is helping dispense with the myth that the scientific conclusions are unsettled.

Discussion questions

1. What do you think is the most significant obstacle to tackling climate change, public scepticism or the difficulty in motivating people to change their behaviour?

[17] Orsekes and Conway also discuss a book entitled *Bad Science: A Resource Book*, which they describe as 'a how-to handbook for fact fighters, providing example after example of successful strategies for undermining science, and a list of experts with scientific credentials available to comment on any issue about which a think tank or corporation needed a negative sound bite', Orsekes and Conway (2011, 6).

[18] Quoted in Shome and Marx (2009).

[19] See, for example, http://www.people-press.org/2014/01/27/deficit-reduction-declines-as-policy-priority/.

2. What's the difference between attributing climate change to natural variability and asserting that we don't know why we are experiencing unusual climate change? (Hint: maybe there isn't a big difference.)
3. Do we have a responsibility to future generations to protect the planet? How much should we be willing to sacrifice to protect the unborn?
4. Which nations and geographical regions are likely to suffer most from climate change? Do we have a responsibility to protect those who live there?
5. Why do you think scepticism concerning climate change persists, and what do you suppose would be the most effective ways to overcome it?
6. Which do you think is most open to reasonable doubt: that the planet is getting warmer, that atmospheric CO_2 is increasing, that the latter is responsible for the former, that the warming trend will continue or that the climate will become increasingly hostile for both human societies and natural systems?
7. Do you think it would make a big difference to public opinion if it became more widely known just how significant the scientific consensus on climate change is? If not, why not?
8. To what extent do you think people are unconcerned about climate change because they regard it solely as an *environmental* issue, and hence not one that will affect people? What would you say to anyone who adopted something like this attitude?
9. Scientific efforts to understand human effects on the environment include studies not just of climate change, but also endangered or threatened species, soil contamination, air and ocean pollution and so on. Do you think that such scientific investigations are particularly vulnerable to outside interference and artificial criticism? Why might that be?

Suggested reading

For good overviews of the state of the science surrounding climate change, see Houghton (2009), Archer and Rahmstorf (2010), Pilkey and Pilkey (2011) and Maslin (2009). Flannery (2001) is less technical, but still informative. Weart (2008) provides an excellent history of the sciences surrounding climate change. Shulman et al. (2012) offer practical strategies for reducing carbon emissions that we can all employ. A very informative website, which provides thorough responses to many sceptical objections to the sciences surrounding climate change and links to peer-reviewed work, can be found at www.skepticalscience.com.

9

Sciences, religion and an intelligently designed controversy?

In 1999 the Kansas Board of Education was involved in what became a very public review of its standards of science education. Up for discussion, and piquing public interest, was what those standards would say about the teaching of biological evolution. Commenting on the case Phillip Johnson would later suggest: 'What educators in Kansas and elsewhere should be doing is to "teach the controversy"'.[1] Johnson was a law professor at U.C. Berkeley, author of *Darwin on Trial*, and co-founder of the Discovery Institute, a Seattle-based think tank that has worked tirelessly to promote what's known as Intelligent Design (ID) theory. The nature of the controversy that Johnson thought worth teaching is not entirely clear, but the idea that we should 'teach the controversy' became an important campaign for the Discovery Institute, a campaign that would soon gain significant momentum. By August 2005, President George W. Bush was convinced, proclaiming that 'Both sides ought to be properly taught' and that 'Part of education is to expose people to different schools of thought'.[2]

The Design Institute's teach-the-controversy campaign is just one chapter in the much bigger story of how evolutionary biology and Creation science have been regarded within U.S. public school systems, a story that traces back to at least the 1920s. And the story of evolution's place in the classrooms is a small part of an even bigger set of issues concerning the relationship between the sciences and religions. Here we can barely hope to scratch even the surface of many subtle, interesting and worthwhile topics. We will consider whether the controversy we were being urged to include in sciences classrooms is a created one, and what attitudes towards evolution and ID are reasonable. We'll start, however, by considering the more general issue of how the sciences and religions are related.

Sciences and religions

It appears to be quite widely supposed that the sciences are straightforwardly in conflict with religion, but what does this actually mean? Many religions appear

[1] Johnson (2002). [2] Quoted in the *New York Times*, August 3, 2005.

concerned principally with worshiping a god, or gods; with moral codes; with describing the best way to lead a good life; with the significance of certain historical events, parables and teachings; with questions about the meaning of life; what happens to us when we die; our responsibilities to other members of our community and so on. Different religions, or denominations, might disagree on some of these issues, but the clashes don't obviously involve sciences.

A better place to look for sources of conflict lies of course in the fact that religions don't always limit themselves to the above roles and questions, but also make claims about when the world was created, how it was created and how god continues to intervene in worldly affairs. As scientific disciplines have unfolded, they have generated, and subsequently accumulated, impressive evidence in support of conclusions that contradict beliefs based on religious texts. A literal reading of the Bible, for example, suggests that the Earth is just thousands of years old, rather than the billions of years reported by modern sciences. In some cases, sciences meet opposition from those who insist upon the infallibility of certain sacred passages, but the response of theists has been far more varied than that of dogmatic resistance. A great many theists seem quite happy admitting that the Earth orbits the Sun and is billions of years old, that species evolve over time and that an important mechanism of evolutionary change is natural selection. The pope, for example, accepts the evolution of biological species under natural selection, as do a great many other theologians, archbishops, bishops and church leaders. These thinkers, and many others besides, grant scientific authority on at least some matters, without abandoning religious beliefs altogether. If the conflict between sciences and religion is supposed to arise from the conflicting conclusions they generate, concerning the age of the Earth for example, then the fact that many religious believers are comfortable admitting scientific authority on such issues suggests that conflict is not the only, and hence perhaps not the best, way of conceiving of the relationship. We'll return to conciliatory attitudes shortly, but it's first worth recognizing another way in which conflict seems to arise.

According to some critics of religious belief, a proper appreciation for the nature and success of the sciences shows us (among other things) that it is irrational to believe something without good reasons for doing so. Thus, insofar as religious beliefs seem to rely on types of evidence that sciences regard as untrustworthy, such as ancient texts, personal revelation and the authority of religious leaders, then religious convictions are judged irrational. Theists respond that religious beliefs are justifiable and that arguing otherwise requires an overly restrictive view of what counts as good evidence or a reasonable belief. The broader debate extends beyond the merits of particular scientific conclusions, drawing us into a much more ambitious project of deciding what kinds of evidence should be taken seriously, and how much significance they should be afforded.

The merits of certain scientific arguments may force us to rethink particular, religiously inspired beliefs, but admitting that the empirical evidence surrounding the age of the Earth is extremely compelling does not entail that all religion is irrational. There are many philosophical arguments advanced in favour of, or opposition to, the existence of a god, the reasonableness of belief in god and what combinations of attributes god can or cannot have. These are arguments that have little to do with when the world and its inhabitants were created, or how. To suppose that sciences settle the philosophical questions is to suppose that scientific investigation is the only way to acquire genuine understanding. In assuming this, then of course the sciences will conflict with worldviews that purport to reveal knowledge via other means. The plausibility of alternative means of generating reliable beliefs is something we can only evaluate by engaging them.

It might sometimes *appear* as if we each must either admit the authority of the sciences but then recognize religions as outdated, naïve and irrational, or retain religious sensibilities and deny at least some scientific claims. However, there are other ways of conceiving of the relationship between science and religion that deserve attention. Presenting the predicament as such a blunt choice is not just misleading, it risks derailing the discussion, by nudging many towards a particular attitude without having attended to the quality of the arguments or considerations being advanced. Public support for teaching Intelligent Design emerges despite almost no grasp for the content of that theory's claims or arguments, much less their merits. Support is instead motivated by religious beliefs and the perceived incompatibility of those beliefs with at least some aspects of evolutionary biology. Ideas and arguments are dismissed by others for no better reason than that they were advanced by theologians.

Various considerations portend a more complicated relationship than the caricature of two worldviews in conflict suggests. For several centuries following the scientific revolution, many thinkers supposed that *because* the universe was designed by an omnipotent and omniscient god, then there must be some rational design and order that diligent human investigation might gradually reveal. Attempts to understand the world were in part motivated by religious convictions. Simultaneously, by attempting to better understand the creation, and thereby exercising our god-given talents of intelligence, curiosity and rationality it was thought that scientists glorified god. Major religions have often directly promoted studies that we would today regard as scientific, sometimes in service to their religious practices. Within many religions, for instance, it is important that festivals, holy days and times of prayer be celebrated or observed at the appropriate times, but calculating when these events should be observed motivated advances in both astronomy and mathematics. Many other scholarly interests were encouraged and supported within religious houses and institutions despite no direct application for religious practice or interpretation. We do a gross disservice to the history of the relationship between sciences and religion if we suppose that it has been one of perpetual conflict.[3]

[3] The most famous historical case of sciences and religion at apparent loggerheads involves the trial of Galileo by the Roman Catholic Church in 1632. Even here, however, the historical details reveal a more complicated story than simple conflict between science and religion. Historians now suggest that the affair is better parsed in terms of the

Even beyond historical considerations, the image of unambiguous conflict is not obvious. Many scientists, respected and active within their profession, are observant of religious doctrines and report belief in god. We saw in Chapter 1 that the problem of demarcation has admitted of no easy solution. Defining religion is similarly challenging. What might be central to some world religions can be almost entirely invisible within others. For any particular religious tradition, disagreements surround the interpretation of certain claims as well as the available responses to theological problems. With the proper definitions of science and religion both contested, debates about their compatibility become more complicated. None of these considerations establish that the sciences and religions should not, ultimately, be understood as locked in an ongoing struggle with one another. There are certainly some religiously inspired beliefs that conflict with some scientific conclusions. Nevertheless, we should at least be curious about alternative ways of understanding the relationship. Ian Barbour is one theologian who has offered a useful taxonomy of perspectives on the relationship that are worth reviewing in brief compass.[4]

The first relationship described by Barbour is that which we've already discussed. Perhaps, admits Barbour, conflict is the most sensible way to conceive of the relationship. Religions make substantive claims about the world we live in, but scientific conclusions contravene them; one must be wrong, so individuals are forced to adopt either a scientific view of the world or a religious one. This attitude has been championed by groups from each side. In some cases, critics regard the opposing worldview as not only misguided but dangerous. Critics of religions cite religious wars, religiously motivated surgical practices like female circumcision and instances of minority groups being abused or oppressed on religious grounds, all as evidence that religions are not just wrong but a significant source of suffering and misery. Sciences, on the other hand, are blamed by some for the development of weapons of mass destruction. Some people fear and distrust technologies like genetic engineering and nuclear power. Others perceive society as falling ever deeper into corruption and sin, a fall they associate with materialistic sciences and which they consequently desire to overturn. The questions of whether sciences or religions are inherently dangerous won't concern us further, but they remind us of how much ground the debates now cover.

Barbour's second means of relating sciences and religions supposes they are *independent* of one another, concerned with different kinds of issues, and appearing to conflict only because each is sometimes extended beyond its proper boundaries. Stephen Jay Gould was a Harvard palaeontologist who wrote many popular books on evolutionary biology. Gould defended the thesis of independence when he described sciences and religions as *non-overlapping magisteria*. He argued that there is no

Church's response to its authority being challenged, rather than a dispute over how we should settle empirical claims.

[4] Barbour (1990).

conflict because there is no 'overlap between their respective domains of professional expertise'.[5] According to Gould, the sciences are concerned with 'the empirical constitution of the universe, and religion in the search for proper ethical values and the spiritual meaning of our lives'. When sciences and religions are properly confined to their respective domains, therefore, conflict between them becomes impossible. The appearance of conflict arises only because some use either sciences or religions to encroach on issues that they are not actually well suited to address.

Gould's view has been much discussed and much criticized. One significant worry with the suggestion is that limiting most major religions to claims about value and meaning renders those religions almost unrecognizable. Consider some familiar claims from Christianity: that the world was created in six days and that humans were there from almost the very beginning; that Noah survived a global flood; that Jesus performed miracles, and was raised from the dead; and that god answers prayers. The difficulty for those who adopt Gould's view is that these are all *empirical* claims. These are not moral claims, nor are they directly concerned with the meaning of human life. They concern, in part, what has happened or what will happen. Admittedly, some of these ideas might be very hard to evaluate, to confirm or disconfirm, but that's beside the point. If we follow Gould in restricting religion to questions of value and meaning, then we can no longer use religion as a source of knowledge about what happened in the recent or distant past, or how god continues to intervene in the world. Christianity is one major religion that looks very different when viewed through the lens of independence. It appears that Gould avoids conflict only by defining religion in a way that excludes the preferred interpretations of almost all world religions.

What's likely to disappoint many theists about the foregoing perspective is its concession that religion is incapable of producing knowledge of our world and ourselves. We can rescue this function by supposing that sciences and religions need neither be in conflict nor entirely independent of one another, but are capable of forging a fruitful dialogue that can provide a fuller understanding than either in isolation could achieve. This is Barbour's third suggestion. The contributions of the sciences to this joint venture are clear. At one time, the role of religion was perhaps to explain why many aspects of the natural world appear to have been exquisitely designed. As we'll see later in the chapter, however, the rise of evolutionary biology has stripped religion of that particular explanatory role. Thus, whether it is reasonable to suppose that sciences and religions can still pursue a mutually beneficial dialogue requires evidence that religions still have contributions to make to our understanding of why things are the ways they are. Insofar as scientific progress and success might seem to proceed entirely independently of religious traditions or assumptions, the prospects of developing this perspective are damaged. A further challenge is to find a role for religion that's more positive than merely filling the gaps in current

[5] Gould (1997).

understanding. As we noted in earlier chapters, all scientific disciplines include unanswered questions. These represent gaps in our understanding, which religion can (and sometimes has) been utilized to fill. Several theologians have argued, however, that god should not be relegated to explaining only what is currently unexplained by the sciences. Part of the worry here is that as sciences advance god is squeezed out of at least some gaps, which undermines the hope that religion could make a positive contribution to our understanding. A mutually beneficial dialogue requires religion to serve as more than a temporary place holder until scientific explanations become available.

Barbour's fourth and final suggestion is that sciences and religions might share a more integrated relationship than merely filling the gaps in our understanding that either in isolation leaves open. Scientific discoveries, for example, might be interpreted as providing critical *evidence* for god's existence. Creationism is sometimes engaged with this kind of project. (However, Creationism is more typically directed towards undermining confidence in prevailing, mainstream sciences; as such, it helps fuel the idea that sciences and religions are actually in conflict.) The major challenge for those who would pursue Barbour's fourth perspective is similar to that which confronts the previous suggestion, which is to say, indicating what advantages are gained by augmenting a purely scientific worldview with one that also integrates religious attitudes.

The purpose of outlining Barbour's four attitudes was not to reach any conclusions about their relative plausibility, but rather to punctuate once more the inadequacy of the dilemma that is often promoted. The choice we confront is not between a scientific perspective and a religious perspective. To those possibilities, we must add a third, that of a reconciliatory perspective that retains aspects of both scientific and religious belief. Crafting such an attitude may not be easy, but ignoring the possibility is intellectually irresponsible. We should also observe that our discussion has assumed throughout that a core function of organized religion is to produce knowledge, but perhaps the function of religion is better conceived in social, psychological or developmental terms. If organized religion benefits the well-being of its practitioners, successfully promotes attitudes that we consider desirable, strengthens community relations and so on, and, furthermore, achieves such ends more successfully than secular organizations, then its function as a source of knowledge could be regarded as less important. (Of course this suggestion raises difficult questions about whether religions *do* fulfil these ends better and, if so, whether they would continue to do so if they're not also understood as a source of knowledge!)

The relationship between sciences and religions is complicated. If I've belaboured this point, it's because public debates surrounding the teaching of evolution and ID are driven – to a large extent – by the impression that evolutionary biology is somehow an attack on religion.[6] What deserves far greater consideration is the possibility that

[6] A 2012 Gallup poll reported that 46 per cent of Americans believe that god created humans in their present form within the last 10,000 years. The proportion who held this belief was considerably higher (67 per cent) among those who attended church at least once a week than it was among those who seldom or never attend church (25 per cent).

evolutionary biology teaches us important lessons about which religious sentiments are reasonable. If it is not obvious why sciences should *ever* influence our interpretation of religious passages, it's worth reflecting on Alan Gishlick's Creation/Evolution continuum, as presented in Eugene Scott's important book.[7] Here we are presented with a range of perspectives, each of which attempts to understand certain biblical passages that are not straightforwardly reconcilable with scientific thought. Gishlick's continuum is comprised of the following positions:

Flat Earthism → Geocentricism → Young-Earth Creationism → Gap Creationism → Day-age Creationism → Progressive Creationism → Evolutionary Creationism → Theistic Evolutionism → Agnostic Evolutionism → Materialist Evolutionism

Starting with Flat Earthism, in Isaiah 11:12 we are told that god 'will assemble the scattered people of Judah from the four corners of the Earth'. A literal reading is problematic – a spherical Earth doesn't have corners. Certainly there is scope for alternative interpretations, but what's more important than discerning how the passage should be read is recognition that the right *interpretation* is needed if the Bible is to avoid the impression that it teaches that the world is flat. (It is worth noting that several further passages also imply a flat Earth and that some individuals have appealed to these verses to defend their view that the Earth really is flat.) The absurdity of a literal interpretation of certain passages places pressure on those who insist on a literal interpretation of the creation stories of Genesis to provide a principled distinction between those biblical passages that should be interpreted literally and those that shouldn't.

According to polls conducted in the late 1990s 18 per cent of Americans falsely believe that the Sun orbits the Earth, with a similar proportion of Germans and Britons similarly confused. I don't know how to explain the origins of this confusion, but historically some have certainly appealed to the authority of holy texts to justify their geocentric convictions. Just as certain verses imply a flat Earth, several others imply a static one. Continuing through Gishlick's list, in 1647 Bishop James Ussher provided a precise estimate for the age of the Earth. His efforts suggested that god began his creation on October 23, 4004 BC. Young Earth creationists don't always commit to Ussher's exact date, but they do reject mainstream sciences by supposing the Earth is just a few thousand years old. Gap creationists read the creation stories of Genesis less literally, supposing that between one day of creation and the next elapsed enormous periods of time. This enables them to reconcile the creation stories with at least some well-established scientific ideas. Day-age creationists take a slightly different view, suggesting that we should not interpret the *day* of creation stories to mean a standard day on Earth as we currently experience it. Perhaps a day for god was billions of years. For day-age creationists Genesis is still an accurate account of the *order* in which

[7] Scott (2009).

things appeared. Progressive creationists doubt even this, preferring the geological record for reliable history, but progressive creationists do suppose that the creation of each species (or perhaps biological *kind*) required a special creative act by god. Evolutionary creationists envisage god as more the master clock maker, who creates things to unfold and evolve according to a plan but perhaps without any need for subsequent intervention. From this perspective biological evolution was god's way of realizing his plan. Theistic, agnostic and materialist evolutionism all admit complete scientific authority on issues of the natural world and its origins, but differ with respect to the possibility of supernatural entities.

One conclusion that Gishlick's continuum makes apparent is that *if* one regards the Bible as a reliable source for information about origins and creation, then either one should accept the Earth is flat (and, quite possibly, additional ideas that look highly implausible by modern standards), or provide a principled means of distinguishing those biblical passages that are to be understood literally from those that are not. Even if one supposes that the Bible might inform our understanding of how and when the universe appeared, we are still required to sort through the many varying interpretations, and defend our reasons for preferring one over the others. Is the Bible a reliable source of information about the Earth's shape, or just its age, or neither of these but reliable with respect to other historical or scientific questions?

Rejecting evolution on the basis of its perceived incompatibility with the Bible prompts quite sensible concerns that individuals are thereby similarly obliged to deny heliocentrism and a spherical Earth. Almost no Christian will be happy with such implications. However, the strategies employed to avoid those implications often provide ammunition for those who would argue compatibility between Christianity and evolutionary biology, via evolutionary creationism or theistic evolutionism. A particularly stubborn stumbling block for contemporary Americans separates Flat Earth views and Geocentricism from Young Earth Heliocentricism. That's to say, many Americans believe the Earth is just a few thousand years old but accept that a spherical Earth orbits the Sun. This is despite the facts, first, that Young Earth Heliocentricism can claim no more scientific credibility than Geocentricism or Flat Earthism and, second, that no principled distinction has been offered to explain why those biblical verses that imply a young Earth should be interpreted literally while those that imply a flat or static Earth should not.[8]

Gishlick's continuum also enforces for us again that the relationship between sciences and religions is complicated. Belief in god is consistent with all except the final two perspectives on his list. The relationship between sciences and religions is

[8] Young Earth creationists are often very keen to deny that the Bible suggests a flat Earth. Why are they so opposed to a literal interpretation? The answer, of course, is their familiarity with the compelling evidence that the Earth is not flat. Scientifically, the issue has been settled. To avoid the appearance of error, these verses must be interpreted non-literally. The problem for YECs, however, is that the evidence for an old Earth, and for the evolution of species, is just as compelling, even if it is less available and requires more effort to understand.

not about choosing between them, but finding an appropriate balance between the authority of each. That there exists the possibility of reconciling science and religion should help us approach questions of biological evolution without prejudice. Unfortunately, the idea that we *should* be willing to listen to scientific evidence on the issue thoughtfully and honestly does not mean that everyone can.

Recognizing that the relationship between sciences and religions is complicated tells us nothing about the merits of evolutionary biology, or ID, or whether the appearance of controversy that surrounds them is genuine or illusory. Furthermore, in one important sense, the relationship between sciences and religions has absolutely nothing to do with ID. Proponents for the latter insist that their theory is *scientific*, not religious. They argue that there is evidence that certain biological systems almost certainly couldn't have evolved through natural processes, hence that the evidence gestures towards a hypothesis of intelligent agency. One issue we'll consider is whether ID can fully detach itself from its apparent religious connotations, but we should start by looking at some of the main ideas in both evolutionary biology and ID.

Evolutionary biology

Evolutionary biology is principally concerned with describing and understanding billions of years of life on Earth. It is a subject that is intimately connected with a great deal of contemporary work in geology, genetics, organic chemistry, biophysics and more. It is both an attempt to describe the changes that species have undergone, across longer or shorter time frames, and to understand why those changes occurred. At its core is the idea that species are not immutable, but undergo change. Over time, these changes accrue. A population of organisms might get a little heavier, or their limbs a little longer, or their plumage a little darker, teeth a little sharper, hearing more acute, hide thicker, petals larger, brighter, flatter and so on. Given enough time, a population might undergo sufficient change that the descendent population comes to look and behave in ways that differ markedly from the ancestral population.

The idea that species evolve into quite different forms is most often associated with the work of Charles Darwin, but he was not the first to entertain such ideas. Erasmus Darwin (Charles's grandfather) and Pierre Louis Maupertuis each independently suggested that species might undergo change. Jean Baptiste Lamarck was a famous naturalist from the early nineteenth century who also argued that species evolve over time. What Lamarck did not anticipate was Charles Darwin's conclusion that today's distinct species are descended from a common ancestor. Bats and whales, for example, are descended from a common placental, mammalian ancestor that lived tens of millions of years ago. Lamarck also explained the mechanism of evolutionary change in quite different terms from Darwin. Darwin's explanation for why species undergo change was his theory of natural selection.

Consider a population of organisms. Within the population there will be variation. Some individuals might have a bigger beak than others, or better eyesight. Some might ward off disease more effectively, or regulate body temperature more efficiently. The number of ways in which individuals from the same population might differ is clearly enormous. Some of those variations will be beneficial within certain environments. In circumstances where food is scarce, those individuals that have a better sense of smell or sight might locate resources more successfully, and thereby perhaps improve their chances of surviving the winter. In some environments, a darker fur colour might provide better camouflage, making those individuals better hunters or less vulnerable prey. Dark-furred individuals within certain environments hold an advantage over lighter-furred members of their population. The capacity to find food and shelter, and to avoid parasites, injury and predators, are all reasons that some individuals will survive longer. Surviving for longer is a good way of improving one's chances of leaving more offspring, although not the only way.

If those advantageous traits are heritable, then the offspring of advantaged individuals are more likely to inherit the same advantages, assuming the environment remains fairly stable. The members of a population that possess advantageous traits, relative to their environment, are likely to have more offspring than other members of the population, and their offspring are more likely to have the same advantageous traits than is a randomly selected member of the population. What follows, if these conditions are met, is that each subsequent generation is likely to be slightly more heavily populated by individuals that possess the advantageous traits. In this way, a population will slowly undergo change and become increasingly well suited to answering the challenges that the environment presents (where the environment includes not only the physical environment, but also other species and members of one's own species). Natural selection is about how *populations* become better adapted to their surroundings.

The theory of natural selection helps us understand how species that look and behave very differently can be descended from a common ancestor. Sometimes members of a population will become isolated from the rest of the group. The environments of the two populations will differ in at least some respects, and sometimes in ways that give preference to different traits. If these populations remain sufficiently isolated from one another, then over time the changes that accrue in each will result in greater and greater divergence. Eventually these may be regarded as distinct species. Natural selection describes a process of small changes to existing tissues, organs and behaviours which accrue over time. Significant evolutionary changes are achieved through a long sequence of stepwise changes, where each change represents some small improvement on (or at least no large retrogression from) the pre-existing form. Perhaps even more significantly, the theory of natural selection also provides the conceptual resources to explain how some of life's most astonishing features can arise through entirely natural processes.

Certain features of the biological world appear to be exquisitely designed. The hummingbird is remarkably well-suited to extracting nectar from certain species of flowering plants. The echolocation system by which bats navigate in the dark, combined with their remarkable flying skills, makes them incredibly effective nocturnal hunters. Some species of orchid have evolved remarkable mechanisms for attracting pollinating insects. Examples of apparently purposeful design abound – and the more closely we study these features, the more impressed we will become.

A particularly favoured example of apparent design has long been that of the eye. Even Darwin recognized the eye as presenting a seemingly acute challenge to his theory, noting that:

> To suppose that the eye, with all its inimitable contrivances for adjusting the focus to different distances, for admitting different amounts of light, and for the correction of spherical and chromatic aberration, could have been formed by natural selection, seems, I freely confess, absurd in the highest possible degree.[9]

However, Darwin quickly moves on to describe how the theory of natural selection might yet be utilized to explain the evolution even of something as complex as an eye.

> Reason tells me, that if numerous gradations from a simple and imperfect eye to one complex and perfect can be shown to exist, each grade being useful to its possessor, as is certainly the case; if further, the eye ever varies and the variations be inherited, as is likewise certainly the case; and if such variations should be useful to any animal under changing conditions of life, then the difficulty of believing that a perfect and complex eye could be formed by natural selection, though insuperable by our imagination, should not be considered as subversive of the theory.[10]

Biologists' understanding of the evolution of eyes has improved enormously over the last 150 years, but retains the basic structure of Darwin's original proposal. We now know that the eye has not just evolved once, but on dozens of occasions across the animal kingdom. Not all eyes are the same. Some species have eyes with no lens, some with just one lens, and others with multiple lenses. The collection of light-sensitive cells of several eye types is cup-shaped, but compound eyes have light-sensitive cells located on a convex surface. One reason for studying other eye types is to gather clues about the types of eye that our ancestors could have had. Our understanding of the evolution of the eye has benefited principally from work in embryology, comparative anatomy and DNA analysis.

The evolution of the eye would have started with the emergence of proteins that are affected by photons, and then cells which capture photons and cause changes in nerves. For the evolution of eyes like ours, a flat patch of light-sensitive cells gradually became concaved to form something cup shaped. The opening of the cup gradually narrowed, so that light was eventually entering through a small aperture, just as with pinhole cameras.

[9] Darwin (1859, 207). [10] Ibid.

The appearance of transparent cells functioned as a rudimentary lens. At each stage, tiny changes could confer an advantage on some organisms, within some environments. The mere ability to detect light is useful for avoiding predators by detecting their shadows. A cup-shaped patch of cells provides information about the *direction* light is coming from. The narrowing of the aperture and formation of the cells can sharpen the quality of the image the eye produces. And importantly, all the changes here described can be understood in terms of a great many very small changes which accumulate over time. Cells become better at sensing light, for example, and lenses become better at focusing. A patch of cells can, over many generations, become increasingly cup shaped. If any small change provides even a small advantage to the organism, then that change has a better chance of spreading through a population.

Zoologists Dan Nilsson and Susanne Pelger sought to estimate how quickly an eye could evolve.[11] In outline, they assumed a patch of light-sensitive cells, and considered how the spatial resolution of that patch could improve through changes in size, shape, closing of aperture and emergence of lens. Nilsson and Pelger were very conservative in their estimates for both the amount of variation within a population and the advantage that small improvements would confer. Concerning the latter, for example, they assumed that for every 101 individuals that survived with an improved eye, 100 individuals would survive without the improvement. Despite assuming what other biologists have agreed were very pessimistic assumptions about the amount of variation in their populations, and the benefits associated with even small changes, Nilsson and Pelger's model suggested that complex eyes, complete with a sophisticated lens, could evolve from a flat patch of light-sensitive cells in less than 400,000 years. In evolutionary terms, this is remarkably fast.[12]

Natural selection does not require that beneficial variations are more likely to appear as they are needed by some population. Move a population into a much colder climate, and the genes responsible for fur thickness don't start mutating in ways that make a coat thicker. Rather the idea is that differences in coat thickness will likely already be among the existing variations within the population and, if so, those with thicker coats might find themselves better able to survive and reproduce. If we assume that coat thickness is heritable, then the population will change via natural selection; each subsequent generation of the population becomes increasingly skewed in favour of thicker coats. Bear in mind also, that while average coat thickness might be increasing within the population, further changes might also be taking place, since the new environment doubtless creates other challenges and thus bestows advantage on other available variations. There are a great many fascinating questions in the details, and in some cases questions that are currently difficult to answer satisfactorily, but hopefully we've at least some sense of what evolutionary biology is about.

[11] Nilsson and Pelger (1994).

[12] As must often be the case in this book, efforts to introduce the state of current understanding can only provide the merest gloss. For a remarkable, book-length account of the evolution of the eye, see Schwab (2011).

The origins of variation within populations, and the mechanisms by which traits are transferred to offspring, are important topics for evolutionary biology. Darwin lacked convincing explanations for either. He gathered a substantive body of evidence that natural populations did possess variation and that traits are heritable, but this didn't help him understand why. Many have noted the irony that Darwin had access to Gregor Mendel's work on heredity, but almost certainly never read the paper. He was not alone. Mendel's work received very little attention in the decades immediately following its publication. It wasn't until 1900 that biologists rediscovered Mendel's work, appreciated its significance and thus began the study of genetics. By the 1930s, further developments in population genetics, palaeontology and a variety of previously somewhat disconnected biological sciences became more unified in what has become known as the evolutionary synthesis. Central to this achievement was the reconciliation of Darwin's theory of natural selection with Mendelian genetics. The evolutionary synthesis provided a general framework for explaining a wide variety of phenomena. Its core elements remain central to the biological sciences today.

Contemporary thought on biological evolution is about species changing over time, but also about common descent and the mechanisms of change. It is integrated within a much wider scientific image that draws on theories and research in genetics, geology, physics and chemistry. Because evolutionary biology is a complex set of ideas, and because its domain is vast, and because it has a strong historical component, understanding the arguments and evidence that lend credence to the overall picture is not straightforward. Distinct claims require distinct kinds of evidence. Furthermore, as we've emphasized, scientific theories are almost never confirmed by one crucial experiment that answers all doubters. There exists, nevertheless, a truly remarkable quantity of evidence that speaks to the enormous plausibility of the basic image that evolutionary biology describes. Evidence emerges from the fossil record, from observing the geographical distribution of species and by observing degrees of similarity between distinct species. Natural selection can sometimes occur quickly enough for us to observe a population undergoing change. Natural selection is invoked to explain why insect populations have become resistant to pesticides and bacteria to antibiotics. The analogy between artificial breeding and natural selection is also instructive.

Breeders select for breeding those members of their existing stock that best exhibit the traits and characteristics that the breeder desires. A dairy farmer may use for breeding those individuals that produce most milk, for example. Because those traits are often heritable, over many generations a breeder can shape their stock to better instantiate the desirable qualities. Environment can play the role of the breeder, not by consciously deciding which organisms should breed and which should not, but in virtue of the fact that it poses challenges to the organisms that inhabit it. As we've seen, as a result of natural variation, some individuals are better equipped to meet those challenges, and thus the environment can be thought of as *selecting* those advantaged individuals for greatest reproductive success.

The plausibility of evolutionary biology receives further support from advances in molecular biology and DNA sequencing. Although there have been some surprises, DNA sequencing reveals that the more closely related two species are, as judged by earlier methods, the more similar their DNA sequences are. Biologist Sean Carroll offers evidence of common descent and natural selection by appealing to the concept of *fossil genes*. If certain traits are no longer relevant to the survival and reproduction of the members of a population, then mutations in the genes that are associated with that trait will accrue. There's no penalty associated with bad eyesight, for example, if you spend your whole life in caves that remain in perpetual darkness. The genes can thus cease to function without organisms incurring costs. DNA analysis of these genes can yield testable predictions concerning a given species' degree of relatedness to other species and its evolutionary history. Not only has evidence accumulated from a variety of directions, but evolution and natural selection are absolutely integral to our growing understanding of all aspects of life on Earth.

What I've described here is the barest, most meagre, paltry sketch of current understanding. Fortunately there are many wonderful popular introductions to evolution that lay out more of the evidence very accessibly. There are also more comprehensive textbooks, which reference many hundreds of additional sources. Thousands of biologists around the world are employing a wide range of techniques on many different fronts to further our understanding of the history of life on Earth, and they have been doing so for in excess of a century. A wealth of additional information is available to those who actively look. The layperson might often, justifiably, admit to not knowing much of the evidence. No one can sensibly claim that the evidence doesn't exists.

Allusions to biological evolution incite unease, fear and anger among many sectors of society. There are surely a variety of reasons why many aspects of evolutionary biology and, by extension, aspects of geoscience, physics and chemistry, are heavily resisted. As we observed, the debate is often presented as if the sciences are incompatible with core religious tenets. Many theists may thereby feel obligated to express scepticism about aspects of these sciences. Having determined to resist scientific conclusions, a kind of confirmation bias will affect an individual's treatment of subsequent evidence and arguments: arguments that are presented as important challenges to the sciences will be regarded more favourably than those offered in its defence. Consequently, however, while religious commitments will often play an important role in an individuals' scepticism, the reasons they give for dismissing scientific conclusions will likely appeal to perceived critical shortcomings in the sciences. As important as it is to briefly mention the positive evidence for evolutionary biology, doing a little to answer some of the most common misconceptions is also sensible.

In some cases the groundwork has been done in earlier chapters. Critics of evolution demand proof or dismiss scientific theories because they can't explain everything, or because they themselves can't see how the theory could answer their personal source of incredulity, or because someone with advanced degrees offers an objection that sounds compelling to the uninitiated, or because it's *only a theory*. However, as we saw in Chapter 2, sciences don't prove their conclusions. Likewise, no theory explains everything, so it would be inconsistent to dismiss one particular theory merely for being explanatorily incomplete. As we emphasized in Chapter 7, our personal inability to answer certain questions doesn't mean that scientists haven't provided compelling answers, and the fact that some well-educated people are unpersuaded of certain ideas is not evidence that they are in possession of arguments that would be convincing to those who better understand the available evidence and theories. Evolutionary biology can, sensibly, be regarded as a scientific theory, but this provides no justification for resisting its core tenets. Scientific theories play a central role in all scientific disciplines, but the concept of a scientific theory shouldn't be confused with that of a conjecture or idle speculation. It is quite normal for scientific theories to be extremely well confirmed by available evidence and thoroughly tested in a variety of ways. They're still called theories, which is why we have theories of gravity, germs and atoms. Theories are improved as they're expanded or refined, to either explain new phenomena or old phenomena with greater precision, but theories are not promoted to the status of *fact* or *law*.[13]

As well as objecting that evolution is just a theory, opponents also portray evolutionary biology as a kind of ideology. Available evidence is often misrepresented, in service to their belief that evolutionary biology is a theory in decline, supposedly lacking a productive research program and facing an embarrassment of unresolved anomalies. Those who remain confident in the theory, it is concluded, rely on faith rather than evidence. Biologists' rejection of Intelligent Design (ID) can then be explained in terms of institutional prejudice and bias, rather than any actual deficiency with ID. Aside from the quantity, quality and variety of evidence that now supports the core ideas of modern evolutionary biology, the image of evolution as ideology is further undermined by the history of the discipline. As Sahotra Sarkar details, the 150 years of biological thought since the publication of Darwin's *On the Origin of Species* has been a history of change.[14] There have been important theoretical developments, and there remain ongoing conceptual debates within the discipline. The conceptual debates are not trivial, but have arisen as important challenges to prevailing scientific attitudes. The community's willingness to entertain revisions to the existing paradigm undermines the idea that evolutionary biology is an ideology. That the participants in these debates agree on many core elements of evolutionary theory undermines the idea that evolutionary biology is in decline. Evolutionary biology is neither

[13] It is worth noting that many regard evolution as a fact, but here they mean something slightly different. *That* species have changed over time is an extremely well-supported fact. Our best understanding of *how* species change, how populations become well adapted to their environment, how speciation occurs and so on appeal centrally to the theory of evolution by natural selection. Depending what we mean, it can be reasonable to describe evolution as a fact or a theory. On either rendering, these labels do nothing to denigrate evolutionary biology.

[14] Sarkar (2007).

stagnating, nor crumbling, nor failing. It is an evidence-based, vibrant and diverse research program, which contributes immeasurably to our understanding of the natural world.

We'll conclude this section by briefly considering efforts to understand why people struggle to accept what scientists have to say about the history of life on Earth. We've noted that religious belief contributes to significant public scepticism, but work in cognitive psychology suggests further reasons why accepting the facts of evolution are hard work for us. For example, studies of young children indicate that we have strong *essentialist* tendencies when it comes to certain categories, including biological species.[15] This means that we're inclined to suppose that all lions, for example, share some common quality and that it is this quality that makes them lions. To have that quality is to be a lion; to lack it is to be something else. Of course, we might be unsure exactly what the quality is, but studies suggest that we reason as if such essences exist nevertheless.

Furthermore, part of our essentialist thinking is that this common quality is inherited by offspring. Essentialism admits that members of the same species can vary in many different ways, but insists that all must retain that defining quality. The idea that a population could gradually change, either by natural or artificial processes, from one species to another is therefore inconsistent with essentialism. Thus, insofar as we are inclined towards essentialist thinking, accepting biological evolution requires us to overcome our natural cognitive penchants. There is no good reason to suppose that a given population can't be modified and altered to an enormous extent, given sufficient time, but accepting biological evolution might be harder for us than accepting other scientific conclusions.

Our failure to comprehend the time scale of evolutionary change is a further hindrance to our acceptance of evolutionary theory. Multicellular organisms first appeared on Earth over three billion years ago (that's 30 million centuries). This gives evolutionary processes a quite staggering amount of time to generate change. Truly wrapping our minds around such durations is perhaps beyond our cognitive capacity. To illustrate this concept of *deep time*, suppose we compress the entire history of Earth into one calendar year, so our planet forms on January 1. The following events unfold approximately as follows:

First life on Earth: March 22
First multicellular organisms: April 30
First vertebrates appear: November 21
First mammals: December 12
Pangaea (the last supercontinent) starts to break up: December 16
Primates appear: December 26
Colorado river starts slicing out Grand Canyon: December 30, 4 pm

[15] Gelman (2004).

Homo sapiens: December 31, 11:54 pm (6 min before year's end)
Domesticated dog lineage splits from wolves: December 31, 11:56:29
Jesus Christ born: December 31, 11:59:45 (less than 15 seconds before year's end)
American Civil war begins: December 31, 11:59:59 (one second of the year remaining)

Alternatively, imagine driving down the highway. You'll drive 500 miles by the journey's end, and that distance represents the age of our planet. The domestic dog is thought to have split from wolves between around 15,000 and 30,000 years ago. In terms of our 500 miles journey this event occurs with less than six yards to go. You drive for 499.99 miles before the split. In those final six yards, from common stock, dogs evolve into all the wondrous breeds that surround us today, from Rottweilers to dachshunds to greyhounds to bulldogs. If such diversity can be achieved over six yards, consider the possibilities when hundreds of miles of road are available.[16] Thus, while from our perspective a hundred years is a long time, it is entirely insignificant from the perspective of life's history on Earth. Any tendency on our part to suppose that we have a reliable intuitive sense for how much change and what sorts of changes can occur over millions of years is certain to lead us astray. We must look for more reliable means of evaluating the kinds and degree of change that can occur, which of course is exactly what scientists from a range of backgrounds have been doing for decades.

Intelligent Design

What's often advertised as an alternative to evolutionary biology, or at least an important correction and supplement, is Intelligent Design theory, but what exactly is ID? Let's start with what it isn't. First, it shouldn't be confused with mere belief in a creator god. There are many who claim belief in god, but have no sympathy for ID. Even proponents for ID make efforts to distance themselves from equating their theory with religion. Officially, at least, ID theorists conclude only that certain biological features are intelligently designed, but they remain silent on who or what these agents are. For example, ID theorists concede that life on Earth might have been cultivated by advanced extraterrestrials.

Second, ID shouldn't be confused simply with a sceptical attitude towards some or all of evolutionary biology. Being unconvinced of some hypothesis, idea or theory is not in itself an alternative. Furthermore, many leading proponents of ID accept much of evolutionary biology and mainstream sciences more generally. Most ID proponents believe the universe is billions of years old, concede that natural selection is capable of bringing about change within species and even that distinct species are descended from a common ancestor. Michael Behe is one of ID's most prominent advocates. He

[16] No analogy is perfect, and a defect with this one is that dogs have evolved under artificial, rather than natural, selection. Nevertheless, as an illustration of the concept of *deep time* I think the example has value.

writes that 'there's no reason to doubt that Darwin had this point right, that all creatures on earth are biological relatives'.[17] That several leading advocates of ID endorse common descent and an old Earth suggests that public support for ID arises despite much ignorance over what ID defenders actually accept.

ID should be confused neither with belief in god nor blanket scepticism towards evolution. What unifies ID proponents is rather their belief that certain biological or cosmological features are the result of intelligent agency, and that there is compelling scientific evidence for this conclusion. (Since arguments based on biology have been promoted far more by ID theorists we'll focus our attention on these.) Appealing to notions like *specified information*, *complex information* and *irreducible complexity*, it is argued that the appearance of design is sometimes evidence of actual design, and that certain biological systems bear the hallmarks of actual design.

For example, in *Darwin's Black Box*, Michael Behe argues that some biological systems are irreducibly complex. It is extremely improbable, according to Behe, that such systems could have arisen through a sequence of stepwise improvements on earlier versions of that system. As we saw, the theory of natural selection requires that complex systems arise largely as a result of small, stepwise improvements. If Behe is correct in asserting that certain biological systems could not have evolved in this way, then these would create an important challenge to contemporary biology.

Behe's argument is neither that some biological systems seem just too complicated to have evolved by natural selection nor merely that we don't yet have an explanation for the evolution of a particular system. Wondrously complicated systems present no challenge to natural selection, if we can explain them in terms of a sequence of incremental improvements on simpler systems. Of course, biologists might not yet have the explanation, but that's neither surprising nor worrying. Biological complexity is everywhere, and new examples of complex novel systems are being discovered regularly. The absence of explanations for the evolution of particular systems is in itself no reason to doubt that explanations by natural selection are possible. Behe seems to recognize this. His argument is intended to be more probative, and it relies very heavily on his concept of irreducible complexity.

Behe defines an irreducibly complex system as one 'composed of several well-matched, interacting parts that contribute to the basic function, wherein the removal of any one of the parts causes the system to effectively cease functioning'.[18] His illustration of a (non-biological) irreducibly complex system is the common mousetrap, a device that in its most familiar form is composed of a baseboard, spring, hammer, catch and bar. The function of such devices is of course to catch mice. Behe's observation is that this function won't be fulfilled if any of those five components are missing. Removing one part doesn't leave us with a slightly less effective mousetrap, but a thoroughly ineffective one. Taking one component of a mousetrap and

[17] Behe (2007, 72). [18] Behe (1996, 39).

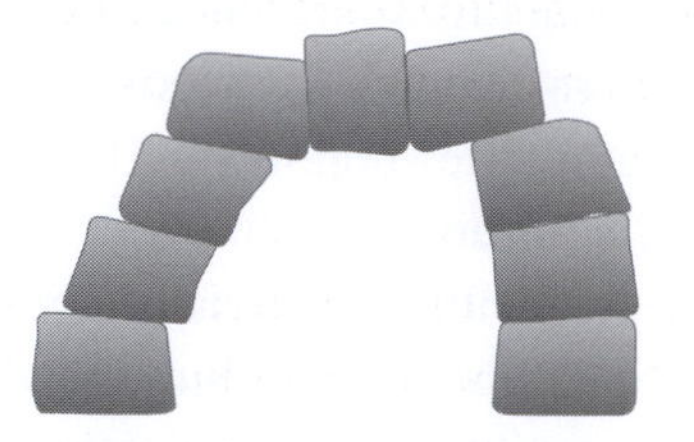
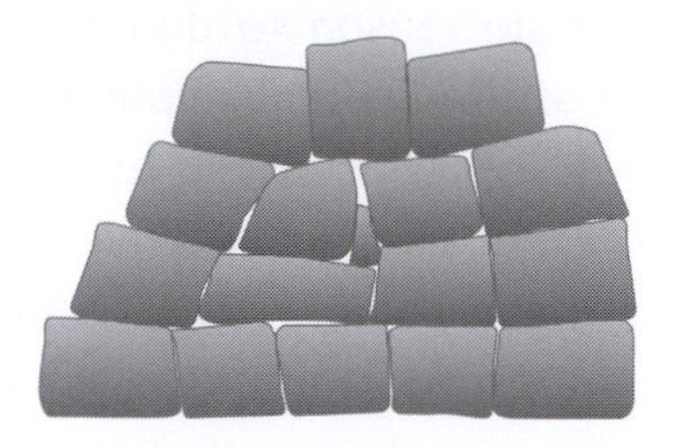

Figure 9.1 The stone arch on the left appears impossible to construct simply by adding one stone at a time. However, a heap, or wall, of stones can clearly arise via such a process and, should internal rocks be removed, perhaps through entirely natural processes, we can produce the stone arch.
(Image adapted from Cairns-Smith (1990).)

introducing a second doesn't give you a better mousetrap. You can't evolve a mousetrap by a sequence of stepwise improvements. Behe claimed to have discovered biological analogues of the mousetrap – biological systems that wouldn't function if one of many interacting parts was removed, hence a system that, according to Behe, either couldn't have evolved through a process of stepwise improvements, or at least where their evolution is so highly improbable that it is effectively impossible.

Scientists and philosophers have identified important deficiencies with Behe's argument and analogy. The most important observation is that biologists have long known of entirely natural processes that are capable of producing systems that satisfy Behe's definition of irreducible complexity. One such process was explained by molecular biologist A.G. Cairns-Smith in the 1980s, by way of analogy with the scaffolding that can be used to help build stone arches. Arches are irreducibly complex: remove any of the stones and the entire arch collapses; you can't build an arch like the one pictured (Figure 9.1 (left)) by adding one stone at a time.[19] However, if you start with a wall or heap of stones and gradually remove the scaffolding, the irreducibly complex arch can be achieved. It can be achieved through an entirely stepwise, natural process. Cairns-Smith argues that biochemical structures can achieve irreducible complexity via an analogous process.

The second route to Behe's conception of irreducible complexity appeals centrally to a possibility that was recognized by Darwin. It concerns the idea that biological systems might be selected for one function at one time, but subsequently selected, or co-opted, for a quite different function. The most frequently offered example is that of bird feathers. Available evidence suggests that feathers were first selected for their thermoregulatory properties. Only much later did they become important for flight, and at points either in between, since, or both, feathers have also been important for purposes of attracting mates, camouflage, diving and much more besides. Such

[19] Cairns-Smith (1990) doesn't use the language of irreducible complexity, so we're using Behe's terminology to describe an argument that Cairns-Smith had actually offered years earlier.

changes in function have no doubt been very common across the biological kingdoms; advances in molecular biology demonstrate that changes in function are widespread at the molecular level.

Suppose, therefore, that a collection of proteins is selected for a given biological function, but where that collection also contributes to a further effect which confers no advantage. As a consequence of an environmental change, let's suppose, the additional effect becomes advantageous. The old function might become redundant, or be fulfilled adequately in other ways. The original collection of proteins might thus be free to evolve through natural selection in service to the newer functional demand. So a system might be irreducibly complex, insofar as removing any of its parts prevents it from fulfilling its *present* function. However, evidence that those parts were once involved in *other* functions ensures that their presence is no mystery, and provides a platform from which we can hope to build plausible explanations for the evolution of those systems in a stepwise fashion.[20]

What both of these responses reveal is that something could evolve entirely through natural processes and yet satisfy Behe's definition of an irreducibly complex system. By co-opting structures that have other functions, and through small incremental improvements to those structures, and the loss of functionally redundant parts, features can arise through entirely natural processes and be irreducibly complex. Irreducibly complex systems are not the reliable indicators of design that ID theorists would like us to believe. Behe was seemingly somewhat aware of these processes when he remarked: 'Even if a system is irreducibly complex (and thus cannot have been produced directly), however, one cannot definitively rule out the possibility of an indirect, circuitous route. As the complexity of an interacting system increases, though, the likelihood of such an indirect route drops precipitously.'[21] Behe has never offered reasons that convinced biologists that the likelihood he mentions here is at all problematic.

Furthermore, when biologists directed their attention to the particular systems that have been offered as examples of irreducible complexity, they began to piece together reasonable explanations for their evolution. Behe and other ID proponents haven't been persuaded by the explanations, but their responses have often involved an important shift in attitude towards seemingly irreducibly complex systems and the capacity of modern evolutionary biology to explain their evolution. Behe's original argument was that irreducibly complex systems were so highly improbable, given chance and natural processes, that they can't be the result of natural selection. ID sympathizers, who have responded to biological explanations for the evolution of those particular systems advanced by Behe, now complain that the available explanations aren't good enough, that biologists haven't gathered enough evidence

[20] For an accessible but much more detailed explanation of how changes in function can explain irreducible complexity, see Miller (2008, chap. 3).

[21] Behe (1996, 40).

to demonstrate that such systems in fact evolved in these ways or that the explanation sketches provided are incomplete in certain respects. However, this transforms the argument back into the uninteresting objection to natural selection that Behe seemed motivated to avoid.

We noted above that our inability to account for every stage in the evolution of every system is unsurprising and unproblematic. In the case of the kinds of molecular structures that Behe draws our attention to, these may have been around for a billion years. But scientists' understanding of such systems is improving. Their confidence in evolutionary biology isn't, shouldn't be and can't sensibly be based on its capacity to fully explain the evolution of everything. It is based on a strong track record of explaining a great deal, on the basis of independently tested assumptions, by drawing on methods from a variety of disciplines, and with more being understood and explained every day.

Ultimately, therefore, Behe's argument from irreducible complexity fails for one of two reasons: if it is offered as an argument about what (almost certainly) could not evolve via natural selection, then it fails because of concepts such as biochemical scaffolding and functional co-option. These explain how systems can be irreducibly complex and nevertheless arise from natural processes. Alternatively, Behe's argument can be reconfigured to assert that because we don't yet know how certain systems evolved, then evolutionary biology must require major revision. It should by now be apparent why this also fails as an interesting objection.

A created controversy?

There are more arguments for ID than Behe's, although the latter does seem to occupy a special place for ID sympathizers. There are also far more technical details than it would be appropriate for us to review, although interested readers are encouraged to consult the suggested reading at the end of the chapter. Furthermore, since most of us lack the necessary training in the relevant disciplines, diving ever deeper into details is unlikely to bring significant enlightenment. It is for this reason that our concern turns now to our three indicators that an apparent scientific controversy is one that has been artificially created.

In this instance the motivation for distorting the state of the sciences is religious belief. Although ID is officially silent on questions about the nature of the intelligence behind the design, this doesn't mean that ID advocates and their supporters aren't advancing views that are religiously inspired and motivated. The task of establishing whether these motives are actual has been made easy by the ID proponents themselves. Jonathan Wells is a leading advocate for ID who revealed that prayer and teachings of the church convinced him to devote his life to 'destroying Darwinism'.[22] William Dembski is another leading ID proponent; he has claimed that 'Intelligent

[22] http://www.tparents.org/library/unification/talks/wells/DARWIN.htm.

design is just the Logos theology of John's Gospel restated in the idiom of information theory'.[23] In 1999, the Discovery Institute outlined its Wedge strategy, and described one of its goals, 'To defeat scientific materialism and its destructive moral, cultural and political legacies. To replace materialistic explanations with the theistic understanding that nature and human beings are created by God.'[24]

In response to observations that ID is motivated by religious convictions, its advocates have responded only that dismissing ideas because of their origins or motives is to commit the fallacy of appealing to motives. But this confuses the reasons that scientists reject ID's conclusions with the reasons that those of us who lack relevant training are justified in harbouring heightened suspicion. Evolutionary biologists claim to have convincingly dispelled the arguments advanced by Behe and others. ID theorists disagree, which is what we would expect in any controversy, manufactured or otherwise. For those of us who lack expertise to evaluate the arguments for ourselves, or are overwhelmed by the number of arguments advanced by ID sympathizers, it is no logical fallacy to be more suspicious of those who have determined to find evidence to support a particular view rather than follow the evidence where it leads. Such determinations will lead to less objective evaluations of the available evidence, and consequently it is not unreasonable for us to regard those same evaluations with more suspicion. Nor does it help that the ID community has insulated itself from the kinds of legitimacy that genuine peer review provides.

There are no reasons to suppose that biologists are motivated in their work by a desire to attack religions or promote atheism. Certainly there are biologists, Richard Dawkins perhaps most famously, who argue that belief in god is irrational and that religion is a significant source of evil. However, this is poor evidence that he pursued scientific research for the purpose of advancing an atheistic agenda. Furthermore, a 2009 poll reported that 51 per cent of U.S. scientists in the biological or medical professions believe in god or a higher power. Given this, and the fact that almost all scientists accept the core components of evolutionary biology, it is not sensible to suppose that biologists at large have an atheist-inspired agenda.

A large number of books aimed at a popular audience either promote ID, challenge evolutionary biology or both. Public debates on these issues regularly take place across the country. Numerous television films and documentaries have been made. National newspapers have featured quotes from many prominent ID advocates. The appearance of conflict within the public domain is clear. However, efforts devoted to alerting laypersons to ID, and thereby creating the appearance of controversy, vastly outstrips the achievements of ID theorists in getting their work published in peer-reviewed journals or convincing the scientific community, as some ID proponents have admitted.[25] The Discovery Institute organized and publicized a petition,

[23] Quoted in Ruse and Pennock (2008, 477). [24] Quoted in Scott (2009, 131).

[25] For example, in 2002 William Dembski described how 'because of ID's outstanding success at gaining a cultural hearing, the scientific research part of ID is now lagging behind', quoted in Forrest and Gross (2004).

intended to discredit evolution and demonstrate the scientific credibility of ID. By 2007 the petition contained more than 700 signatures. However, a large proportion of signatories had no training or background in the biological sciences. Furthermore, as historian of science Ronald Numbers noted, 'Though the number may strike some observers as rather large, it represented less than 0.023 per cent of the world's scientists.'[26]

Finally, we've listed several popular objections to evolutionary biology, and ID proponents have certainly helped promote them, among other objections. Admittedly, ID proponents have advanced their own positive arguments so it wouldn't be entirely fair to say they are concerned solely with raising doubts. However, their efforts to advance positive theses have failed to convince those who are competent to judge and have failed to generate any interesting new results or research programs. Reflecting back on Miriam Solomon's framework, ID has enjoyed no empirical success, and hence the scientific community's rejection of ID is entirely appropriate.

Two of our three indicators point strongly towards a created controversy, and one points weakly in the same direction. Any public perception that there exists a genuine scientific controversy concerning evolutionary biology and ID, and Creation science more generally, is entirely illusory. Those most qualified to judge the relevant sciences are almost unanimous in their approval of the core tenets that evolutionary biology describes. The appearance that the science is unsettled is the result of persistently promoting arguments that have been discredited, distorting scientific evidence to magnify uncertainty and preying on the hopes of a public that is uncomfortable embracing certain facts about the world.

Those who continue to defend ID are forced either to twist these facts or grouse that scientists either ignore the arguments or are intolerant because they're motivated by atheist and materialist prejudice. However, the central ideas of ID have received enormous attention from biologists. We shouldn't confuse the idea that scientists have considered, evaluated and rejected certain arguments or conclusions with the suggestion that scientists are ignoring those ideas. ID advocates have failed to provide any good evidence that the failure of their ideas is a result of institutional bias. The idea that ID is rejected on the basis of materialist prejudice is not plausible.

Having recognized that the apparent controversy is contrived, not genuine, then the case for teaching the controversy collapses. There is no scientific controversy. There might be some very difficult questions orbiting the same vicinity. Is ID a science? Do the sciences presuppose, or imply, metaphysical naturalism (about which, more below)? How much of science education should be directed towards describing shortcomings, anomalies and genuine controversy surrounding well-established scientific research? There are also exciting questions about the details of evolutionary biology, as with any discipline. How and why did multicellular organisms first

[26] Alexander and Numbers (2010, 328).

appear? How did the chameleon's ability to change colour evolve? Why do some species split more than others? Why is eusocial behaviour so much more common in insects than mammals? How does the mantis shrimp generate so much power in its punch? Only the limits of our collective imagination restrict the number of questions that we might ask of the biological world. However, not one of the questions I've posed in this paragraph provides any basis for introducing ID into science curricula.

There are no good scientific reasons to teach ID. Anyone who presses for its inclusion is either deeply misguided about the merits of the idea or motivated by reasons that have nothing to do with promoting a rigorous and sensible science education. The only justification I can see for even mentioning ID within a science classroom is if it is part of something like the following statement:

> Intelligent Design proponents purport to have found evidence that certain biological systems could not have arisen (or very probably did not arise) through natural processes. It is then inferred that those systems must have been designed by an unidentified but intelligent agent. These ideas have attracted significant attention from some members of society, as well as portions of the media. The scientific community is unambiguous in its evaluation. The arguments are weak. The evidence for the central claim is non-existent. There is nothing of scientific value within ID. Time is precious, so we will not be wasting more time on the topic. It was once believed that the Sun orbits the Earth, that smoking cigarettes was good for your health, and that germs had nothing to do with disease. It would be irresponsible of educators to waste time discussing all these ideas. There are not better reasons for discussing Intelligent Design.

Conclusions

Contrary to popular opinion, evolutionary biology is a collection of extremely thoroughly researched and well-evidenced ideas. Its acceptance among biologists, and scientists more generally, is almost universal. The consensus is a product of neither institutional failure nor widespread close-mindedness. Rather, it is based on an enormous body of supporting evidence and a research program that continues to deliver the goods. Evidence has been accumulating for over a century and a half. It comes from a variety of sources and academic disciplines, using a wide variety of methods, techniques and styles of reasoning. Rejecting the central claims of evolutionary biology is no more sensible than rejecting heliocentricism, the germ theory of disease or modern theories of gravity. These claims will no doubt surprise some people, but any surprise is itself evidence only of how successfully the appearance of controversy has been created.

Our personal sense of incredulity is not an objection to the evidence and arguments that favour the central components of evolutionary biology. That we may have unanswered questions is not good evidence that answers are not available for those who actively look. The fact that evolutionary biology can't explain everything is not a reason to reject the many detailed, compelling and comprehensive explanations it can offer for an incredibly wide range of phenomena. Of course, the fact that there is no scientific controversy surrounding biological evolution leaves unanswered a great many questions we may have concerning the relationship between sciences and religions. Religious thought has branched in a great many different directions. Depending which branches we attend to, we are likely to find very different attitudes towards scientific knowledge. Idly dismissing well-established scientific opinion is intellectually irresponsible. The challenge for theists – which a great many do accept – is to find a plausible and intellectually credible means of reconciling religious thought with scientific progress.

Finally, a distinction sometimes drawn in discussions of science and religion is that of methodological naturalism versus metaphysical naturalism. The basic idea of metaphysical naturalism is that nothing exists beyond that which can be studied by the sciences. Metaphysical naturalists thus deny that there is anything deserving of the name *supernatural*. Methodological naturalists, by contrast, are silent on questions about the existence of non-natural entities but are committed to finding natural explanations for our observations and experiences. There is ambiguity and disagreement surrounding each of these concepts, but it is a useful reminder that the relationship between sciences and religions need not be conceived as one of fundamental conflict. Even if the sciences are committed to upholding methodological naturalism, the status of metaphysical naturalism would remain open for discussion.

Some ID theorists have challenged methodological naturalism. William Dembski, for example, objects: 'So long as methodological naturalism sets the ground rules for how the game of science is to be played, intelligent design has no chance of success.'[27] Dembski supposes that methodological naturalism has become scientific dogma. 'By defining science as a form of inquiry restricted solely to what can be explained in terms of natural processes, the Darwinian establishment has ruled intelligent design outside of science.'[28] But Dembski distorts matters when he claims that methodological naturalism is dogma. Even if there are *some* scientists who refuse to countenance non-natural explanations, it doesn't follow that such explanations *wouldn't* be widely adopted *if* compelling reasons for doing so were presented. Earlier chapters have shown that, historically, scientific methods have been subject to revision and replacement. Even deeply held metaphysical beliefs have been abandoned as a result of scientific progress. If there were convincing reasons to introduce non-natural explanations, then they would be adopted, or at least ID theorists have provided no good reason to suppose that they wouldn't. Methodological naturalism is not dogma. Methodological naturalism is adopted by scientists because it works, and because no-one has provided a compelling reason to augment it. The problem for ID is not institutional dogma. The problem for proponents of ID is that they wish to introduce a new methodology, but they have demonstrated neither a deficiency with existing methods nor any merit with the new.

[27] Dembski (2002, 119). [28] Ibid. p. 117.

Discussion questions

1. Gould's attitude towards science and religion has been criticized because religion can't sensibly be stripped of all empirical claims, but also because science can contribute towards our understanding of ethical issues. Do you think science could help us answer questions about morality?
2. If experience and the sciences are more reliable authorities for questions about the Earth's shape and position within the universe, are there reasons to doubt it is a more reliable source for questions about the Earth's age and the evolution of species?
3. To what extent do you think public scepticism towards evolutionary biology is a product of the appearance that there exists a genuine scientific controversy?
4. To what extent do you think public scepticism towards evolutionary biology is a product of gaps in our understanding? Are these good reasons for scepticism?
5. To what extent do you think public scepticism towards evolutionary biology is a product of a perceived absence of evidence for the central claims? Where does the idea come from that no evidence exists?
6. To what extent do you think public scepticism towards evolutionary biology is a product of people not *wanting* to believe it?
7. What are the main strengths and weaknesses of Barbour's ideas of Conflict, Independence, Dialogue and Integration?
8. Are religions dangerous? Are sciences dangerous?

Suggested reading

Fortunately there are a great many very accessible introductions to evolutionary biology aimed at a general audience. Among my favourites are Carroll (2007), Coyne (2009), Dawkins (2010) and Shubin (2009). There are also good textbooks that will provide readers with a better sense of the enormous amount of research being conducted and the thousands of research papers being produced, for example Futuyama (2013). Several books have been written describing the shortcomings of Intelligent Design. Miller (2008), Sarkar (2007), Scott (2009), Shanks (2007) and Brockman (2006) are good examples. Ruse and Pennock (2008) tie together analyses of Creation science with discussions of the demarcation problem; the book includes a nice exchange between philosophers Larry Laudan and Michael Ruse. Isaak (2007) describes and discredits a huge variety of arguments and claims offered in defence of Creation science but answered by subsequent peer-reviewed scientific research. Numbers (2006) is a first-rate history of Creationist science. For helpful discussions of the relationship between science and religion more generally, see Dixon (2008). Ferngren (2002) also includes nice papers.

10

Issues of public health: AIDS, autism and GMOs

If there are secrets to long life and good health, then most of us would wish to know them. If there are products or treatments, supplements or procedures, that would cure our illnesses or ameliorate our symptoms and suffering, then many of us would spend liberally to avail ourselves of the benefits. If our friends, siblings, parents, spouses and, perhaps most of all, our children are confronted with chronic illnesses, severe and considerable daily challenges, acute pain or discomfort, then we will be highly motivated to pursue avenues that promise relief for those we love. Our tendencies here are natural and appropriate. They also make us vulnerable to exploitation.

There is a long history of people who have peddled elixirs and assured customers that their products will cure incurable conditions, prolong life, rejuvenate, revitalize and vivify. Of course, such miracle products come at a price, but the benefits promised are often sufficiently alluring – and potential customers often sufficiently desperate – that collectively we continue to spend billions of dollars on products that can claim, at best, highly questionable therapeutic value. In fact, not only do we part with sums of money we can ill afford, but in some cases we subject ourselves to procedures that are more likely to induce harm than benefit. The claims made by the manufacturers of these products may be nonsense, but if they are dressed in enough technical jargon, promoted by individuals with enough charisma, recommended by celebrities of sufficient stature, given credibility by media outlets that should know better and align closely enough with common wisdom, then the nonsense will often go unnoticed.

There is little doubt that we could all profit from adopting a more critical and careful attitude towards the purported virtues of various expensive cosmetics, herbal supplements, revolutionary diets or so-called complementary and alternative medicines.[1] Nevertheless, the promotion of products through dubious or absurd

[1] The widespread use of the terms *complementary medicine* and *alternative medicine* is itself regarded by some as problematic. The basic worry is that if we lack any reliable evidence that those practices which are described as complementary or alternative are actually effective, then they shouldn't be described as *medicine* at all. Goldacre (2010) is a wonderful exposition of many instances of bad science within the medical and health professions.

claims doesn't typically escalate to the level of creating a scientific controversy. We'll retain our focus on ostensibly created controversies, so this chapter will be concerned with instances when mainstream sciences have come under more direct attack and where the promotion of alternative treatments enters the narrative only more peripherally.

In the next two sections, we'll consider cases from the medical sciences. The first concerns the idea that HIV causes AIDS, a conclusion that has long been accepted by physicians but has been persistently challenged by AIDS denialists. In the second case study we'll review recent fears that childhood vaccinations can cause autism. In both cases our objective is to assess whether the appearance of controversy is indicative of a genuine controversy surrounding our understanding of these issues.

AIDS denialists and critics of modern vaccines are often accused of being *anti-science*, an accusation that is also often levelled at those who oppose the development, cultivation and sale of genetically modified foods. We'll evaluate that assessment towards the end of the chapter. As with the preceding two chapters comprehensive reviews of these issues are beyond our scope. The hope is instead to utilize the distinction between genuine and created controversies for purposes of framing an understanding of these issues, and thereby guarding ourselves against certain kinds of mistakes that are both prevalent and consequential within discussions and presentations of health-related issues.

HIV and AIDS

In the early 1980s, in Los Angeles, California, various medical conditions were being observed in unusually high numbers, conditions that are associated primarily with people who have a compromised immune system. Many of those affected were homosexual men, but heterosexuals and haemophiliacs were also being diagnosed with these opportunistic infections. A common cause was almost certainly responsible for such a prolific number of cases of acutely weakened immune systems. The Centers for Disease Control named the underlying disease Acquired Immune-Deficiency Syndrome, or AIDS. Within a few years laboratories in the United States and France had identified HIV (Human Immunodeficiency Virus) as the cause of AIDS, and efforts were begun both to better understand HIV and the progression of the disease, as well as to develop treatments and hopefully a cure.

Unlike most of the viruses we battle, HIV is a retrovirus. AZT is an antiretroviral drug that was developed in the 1980s. It has been shown to prolong the lives of HIV patients. By the late 1990s, HIV patients were being prescribed a combination of antiretroviral drugs that disrupt different stages of HIV's progression. Utilizing a cocktail of drugs in this way is known as Highly Active Antiretroviral Therapy (HAART). The medical profession's understanding of AIDS/HIV remains incomplete, but much can be very confidently asserted, is justified by thousands of studies and accepted by thousands of AIDS scientists. HIV exists, and its infection causes the systematic collapse of the immune system. HIV is transmitted from one person to another via one of several body fluids. The most common forms of transmission are unprotected sexual intercourse, blood transfusion, including via needles that have been contaminated with HIV-infected blood, and breast milk. It is also abundantly clear that antiretroviral therapy is extremely successful. AIDS scientist, Paul Volberding, wrote recently in a leading medical journal that:

> Advances in understanding of HIV biology and pathogenesis, and in application of that knowledge to reduce morbidity and mortality, rank among the most impressive accomplishments in medical history. No example since penicillin rivals the development of antiretroviral drugs in controlling a previously fatal infection.[2]

Despite the near universal consensus among physicians and medical professionals, there exist several groups that vociferously and persistently challenge prevailing scientific wisdom. Some deny the existence of HIV. Others admit that the virus exists, but regard it as harmless and thus deny it causes AIDS. Some dispute the efficacy of available treatments and even suggest that the antiretroviral drugs may cause AIDS. Some attribute AIDS to poor nutrition, recreational drug use or lifestyle choices. Collectively those who express acute scepticism concerning AIDS science are known as AIDS denialists. The efforts of such groups to promote their ideas and influence the behaviour of others are not without consequence. The website aidstruth.org provides a list of approximately two dozen prominent AIDS denialists who have died with symptoms or illnesses that are associated with late-stage HIV. It's hard to know how many others have been convinced to forgo antiretrovirals and have died prematurely as a result, but have kept their rejection of mainstream science largely to themselves, such that the role of AIDS denialism goes unnoticed. Events in South Africa, at the beginning of this century, provide a more staggering illustration of the damage of denialism.

Thabo Mbeki became president of South Africa in 1999 and convened an AIDS Advisory Panel the following year. Of its thirty members, almost half were AIDS denialists, despite the fact that within the medical professions denialists represent a quite negligible proportion. Mbeki subsequently appointed a health minister, Manto Msimang, who was herself an outspoken denialist. Through new policy and by spreading confusion about HIV, the importance of HIV testing, the efficacy of available treatments and the motives of the pharmaceutical companies that manufacture antiretrovirals, Mbeki and Msimang were responsible for extraordinary and unnecessary suffering. One study concluded that South African policy during the early years of Mbeki's presidency resulted in 330,000 preventable deaths and 180,000 avoidable HIV infections. These conclusions must be allowed time to sink in. The

[2] Quoted in Nattrass (2012).

deaths were preventable. The new infections were avoidable. There were no good reasons for Mbeki and Msimang to ignore prevailing medical wisdom. They were forewarned that their actions invited disaster and that the ideas of those denialists they were relying upon had been thoroughly discredited. They chose to listen to a tiny minority of scientists, who were far less qualified in the relevant theories and practice.

Nicoli Nattrass is a South African economist and an authority on AIDS denialism. She recommends that laypersons keep in mind two crucial facts when reflecting on the issue of AIDS denialism. First, clinical trials have demonstrated unequivocally that antiretroviral drugs lengthen the lives of HIV patients. Second, AIDS scientists have a good understanding of *why* antiretrovirals work, an understanding that is based on the biology of HIV and the mechanisms by which it invades human cells. Confronted with these claims, the case for any form of AIDS denialism crumbles. We must conclude either that there exists a conspiracy within the medical professions of truly breathtaking proportions, such that no part of mainstream AIDS science can be trusted, or we must concede that any residual uncertainties within the sciences are peripheral to the core issues. The conspiracy theory is not plausible. There are too many AIDS scientists, all in broad agreement about the basic results. Uncertainties are expected, which are topics for ongoing research, but in themselves they do not cast doubt on those core conclusions, that antiretrovirals are effective and that AIDS scientists know why they're effective.

The fact that denialist groups hold stubbornly to discredited hypotheses and speculations is no reason to doubt mainstream HIV sciences. Very few scientists or physicians are AIDS denialists, and those that are lack expertise in HIV research. Even denialists concede that the majority view among AIDS scientists is close to a complete consensus, but denialism lives on. It has had tragic consequences, but the existence of AIDS denialism is not very widely appreciated. It is nevertheless instructive to consider why some HIV patients are willing to risk their lives, and in some cases the lives of their children, for no good reason.

Nattrass identifies four roles within the AIDS denialist community that help sustain the illusion, and may even be necessary for denialism to persist. These are the hero-scientist, cultropreneur, praise singer and living icon. Hero-scientists give the cause scientific credibility because they deny at least some key aspect of the relevant mainstream science while possessing a scientific pedigree. Cultropreneurs promote alternative therapies and treatments and thus stand to profit from the perception either that major pharmaceutical companies can't be trusted or that the science surrounding HIV and AIDS is worryingly uncertain. These individuals are willing to invest in denialism as part of their business model. Praise singers are journalists or documentary makers who help spread misleading information. Living icons provide the public example of what can supposedly be achieved by ignoring the recommendations of the health professionals and pursuing alternative remedies and treatment.

As Nattrass also observes, however, the hero-scientists of AIDS denialism fail to secure funding for their projects, fail to advance our understanding of the disease, fail to convince their peers that their ideas are even worth publishing in reputable medical journals and yet continue to promote the same arguments that the overwhelming majority of researchers have long since recognized as flawed. Cultropreneurs can't provide more than highly dubious anecdotal evidence that their supposed therapies are effective and in some cases have been prosecuted for propagating misleading information. Praise singers are storytellers, not reliable sources, and the living icons die of AIDS-related diseases, although their supporters may insist that the deaths are not related to HIV or AIDS. Nattrass's analysis of the denialist communities is instructive. A veneer of scientific credibility, combined with financial backing, efforts to publicize discredited ideas, and living exemplars of alternative methods, can give rise to the dissemination of dangerous lies that are judged plausible within some communities. We might nevertheless still find it surprising that individuals could be convinced of the fabrications.

Seth Kalichman is a social psychologist who has written extensively on HIV prevention and treatment.[3] He describes denial as a normal and healthy response to information that initially we find too painful or difficult to accept. Denial is an important coping mechanism. It is to be expected that some who are diagnosed with potentially fatal illnesses would be susceptible to the idea that the sciences are flawed and that the patient has no cause for concern. Kalichman also describes research which has shown that patients who ask challenging questions of their doctors survive longer. Denial and scepticism are each appropriate and potentially beneficial traits. However, as Kalichman continues, when either of these stances becomes sufficiently entrenched, patients will adopt increasingly irrational and self-harming behaviour.

Communities that promote AIDS denialism inflate denial and scepticism to dangerous levels. Prevalent means by which these ends are achieved include promoting suspicion of the multinational pharmaceutical companies that manufacture antiretroviral drugs and cherry-picking those particular studies and expert opinions that promote their agenda but have been refuted within the scientific literature. Individuals are encouraged to trust their own ability to evaluate the evidence over the collective capacity of an entire scientific community. We can expect cognitive biases to further blind patients to more sensible conclusions. For example, as individuals avail themselves of more denialist literature, partake in denialist online chatrooms and blogs and view denialist-funded documentaries, then the repetition of certain ideas will itself induce greater confidence in the denialist position.

These are all examples of themes we have discussed in earlier chapters. They should help us better appreciate why denialism persists. The story of AIDS denialism also cautions us to remember that each of us harbours subconscious desires and

[3] Kalichman (2009) is an excellent resource for many important questions concerning AIDS denialism.

tendencies, and that these compromise our ability to evaluate evidence. If denialists' gravest error is ignoring without good reason the wisdom of experts, then consistency requires that we check ourselves very carefully on those occasions when we find ourselves tempted to dismiss expert, consensus conclusions. We will return to AIDS denialism again shortly, to consider its relationship to our concept of a created controversy, but not before reviewing another important medical issue which has suffered considerable scrutiny from the wider public and media.

Vaccinations and autism

Smallpox is a nasty disease. Symptoms become apparent only a couple of weeks after infection and include backaches, headaches, nausea and vomiting. A few days later, rashes appear that soon cover the entire face, as well as neck, hands and feet. The rashes start as red lesions, but quickly become filled with pus. In many cases, these pus-filled sacs expand and eventually burst, releasing an unpleasant odour. Some forms of smallpox involve significant internal bleeding. Those who survive the disease are often left blind and significantly scarred across the face. The disease has been around for thousands of years, with some evidence suggesting it might have evolved several tens of thousands of years ago. Prognoses differ from one form of the disease to another, but across all forms it is fatal in approximately 30 per cent of cases. Devastating epidemics have swept through Europe, Asia, Africa, the Americas and Australasia. Some estimates suggest that 500 million lives were lost to smallpox during the twentieth century. In the early 1950s, tens of millions were dying from smallpox every year.

Fortunately, smallpox is now a thing of the past. Globally there have been no reported cases across the last three decades. Pivotal to the disease's eradication were the development and distribution of smallpox vaccinations. Many other diseases, like polio, pertussis and diphtheria, have been eliminated from many regions, or have at least become far, far less common. Vaccinations have saved tens of millions of lives, and millions more have been spared paralysis, birth defects and brain damage. The success of many vaccination programs ensures that we are all far less aware of the symptoms of these diseases and the horrors of epidemics. The extraordinary benefits of vaccinations are easily forgotten when the consequences they are designed to combat are unfamiliar to us.[4]

Public concerns about vaccine safety have been around for a long time, but, particularly since the late 1990s, those concerns have coalesced increasingly around the idea that vaccines cause autism. One of the seminal events in the rise of this story was the publication of a paper in 1998 in *The Lancet*, a prestigious medical journal. The lead

[4] There are many excellent resources for reliable information about the history of vaccines, vaccine safety and the recent scare surrounding vaccines and autism. Mnookin (2012) and Offit (2010) are excellent texts. Useful online resources can be found at cdc.gov, nhs.uk, who.int and vaccinesafety.edu.

author was Andrew Wakefield. In the paper, Wakefield and his co-authors suggested a possible link between autism and the MMR vaccine. The vaccine is administered via a single injection, but protects against mumps, measles and rubella by combining vaccines against all three diseases. Wakefield's suggestion, which extended beyond the scope of the published paper, was that the measles component of the MMR vaccine travelled to the intestine, where it caused infection and inflammation. This part of Wakefield's theory explained why the patients he had examined suffered from inflammatory bowel disease. Because the intestine was now damaged, according to Wakefield, it would 'leak' and, in particular, could allow harmful proteins to enter the bloodstream from the bowels. If those proteins – which Wakefield admitted he had not yet identified – were migrating to the brain, they could cause autism.

The list of problems with Wakefield's work has become a long one, and concerns not just methods and analysis, but also his motives. Immediately, there were puzzles: cases of autism were first reported in the 1930s, long before the MMR vaccine became available; the behavioural problems associated with autism were usually reported as having *preceded* bowel problems, an anomaly for Wakefield's hypothesis. The published paper was based on studies of twelve autistic children, an incredibly small sample, given the millions of such vaccinations that have been administered worldwide. Far worse was that subjects had been referred to Wakefield *because* they were both autistic and suffered with acute bowel issues, and Wakefield was known to be interested in connections between the two.

The only evidence for connecting autism and bowel problems with the MMR vaccine was parental recollection that the first signs of behavioural changes occurred shortly after their children had received the vaccinations. Wakefield had offered no evidence that the measles virus persisted in the bowels of his subjects – it was the inflammation alone that motivated the entire theory, coupled with parents' suspicion. Several years later, the London School of Medicine's dean of research described Wakefield's paper as 'probably the worst paper that's ever been published in the history of [*The Lancet*]'.[5]

The mere suggestion that MMR might cause autism, despite being based on such flimsy evidence, nevertheless motivated further investigation. Scientists in Finland, Boston, California and Denmark studied the medical records of millions of children for signs that receiving the MMR vaccination increased that risk of autism. The correlation wasn't there. Available evidence spoke heavily against any connection between MMR and autism, but the idea didn't go away. One lab claimed to have discovered the measles virus in the intestines and spinal fluids of many autistic children, but only rarely in nonautistic children. Other labs found no such evidence, and independent examiners later determined that the lab which claimed evidence supporting Wakefield's hypothesis was unaccredited and unreliable.

[5] Quoted in Mnookin (2012, 111).

The problems extended beyond anomalies, an acutely speculative explanation that lacked supporting evidence at every step and a deeply problematic selection bias. After several years, Brian Deer, an investigative journalist, discovered that Wakefield had been paid by a law firm. That firm was representing the parents of some autistic children who were seeking damages from pharmaceutical companies that manufactured vaccines on the grounds that these were responsible for their children's autism. In some cases, these were the parents of those children who were described in Wakefield's study. Deer also discovered that several months before the publication appeared Wakefield had filed a patent for a measles vaccine that would be administered independently of vaccines for mumps and rubella. This is, of course, exactly the kind of product he would later recommend, without ever revealing the conflict of interest.

The Lancet would later retract the paper. Ten of Wakefield's twelve co-authors would issue a retraction of the paper. Of the 1998 research, they asserted that 'no causal link was established between MMR vaccine and autism as the data were insufficient'.[6] The journal's editor later admitted that if the conflicts of interest had been known, the paper would not have been published. The UK's General Medical Council described aspects of Wakefield's work and conduct as dishonest, misleading and irresponsible, and as having brought the medical profession into disrepute. Wakefield has since been struck from the medical register.

Wakefield's response has been to paint himself as simply an advocate for children's welfare, a doctor willing to listen to parents' concerns and suspicions, someone unwilling to expose children to the risks he claimed to have identified with the MMR vaccine. He largely ignores the grave problems with his study. Several journalists for the national media, however, were not only willing to report Wakefield's conviction that the MMR vaccination was a possible cause of autism, but in some instances portrayed him as a hero-scientist who was willing to stand up to the medical establishment and greedy pharmaceutical companies. Media sympathy for Wakefield's theory had predictable consequences. The proportion of children receiving the MMR vaccination dropped. The number of cases of measles increased and, tragically, in a small number of cases the disease proved fatal.

The story of Andrew Wakefield's claim contains many details, some of which are no doubt open to debate. Wakefield's supporters continue to quibble over some of them. For purposes of understanding the more general issue, however, we should be sure not to lose sight of two points: first, massive epidemiological studies overwhelmingly suggest that receiving the MMR vaccine has no effect on the risk of developing autism; second, not one of the links in Wakefield's account, from MMR to bowel infection and inflammation, to leaky gut, to dangerous protein escape, to migration to the brain, to autism, has ever received any credible supporting evidence.

[6] Quoted in Offit (2010, 41).

Wakefield related autism to the measles component of the MMR vaccine. MMR has never contained thiomersal (spelled *thimerosal* in the United States), a compound that contains mercury and which has been widely used as a vaccine preservative, but it was thiomersal that would come under enormous scrutiny during the early years of the new century. Evidence of a connection between thiomersal and autism would have provided no support for Wakefield's hypothesis, but, regardless, this putative connection has been just as effectively discredited.

Thiomersal was first used in vaccines in the 1930s and has since been widely used as a preservative in vaccines and other products. Early studies suggested that in addition to being cheap to make, thiomersal had no side effects, was an effective weapon against a range of microbes and didn't seem to compromise the function of the vaccine. The vaccination schedule has grown significantly during the twentieth century and for many decades little attention was paid to the effects of introducing more and more thiomersal into a child's system. New vaccines were each evaluated on their own merits. In 1999, efforts were made to assess any risks. The motivation was largely lack of relevant data, rather than an identifiable cause for concern. There were reasons to suspect that thiomersal posed no risk to human health, but some individuals sought greater assurance, and since it would take time to conduct the desired studies, some changes were introduced to the recommended vaccine schedule. None of those who pushed for more research, or changes in policy, ever mentioned autism in particular. However, the reports that were issued that summer piqued the interest of parents, some of whom began searching online for more information about mercury poisoning (since thiomersal contains mercury). In their opinions the symptoms associated with mercury poisoning were similar to those of their autistic children. Suspecting that these similar effects must have similar causes, armed now with the information that some vaccines contained a mercury compound, these parents now came to believe that thiomersal caused autism.

There are many dimensions and important questions surrounding the thiomersal scare – political, psychological and social – but if our concern is to reach reliable conclusions, then the central issue is straightforward: is there any evidence that thiomersal causes autism? The answer is a resounding *No*. Neurologists know that the symptoms of mercury poisoning are quite different from those of autism. There is no evidence that autistic children have more mercury in their bodies. Two types of epidemiological study have been conducted to investigate the relationship between thiomersal and autism; each showed that thiomersal does nothing to increase the chances of developing autism. One type of study takes advantage of the fact that thiomersal was removed from vaccinations in many countries. It is thus possible to evaluate rates of autism diagnoses before and after these policies were introduced. Studies from several countries reveal that removing thiomersal made no statistical difference to the number of reported incidents of autism. A slightly different type of study reviews medical records to see whether autism is more prevalent among those who received vaccinations that included thiomersal. Studies in California, Sweden, Denmark, Canada and the UK have reviewed thousands of records and found no increased risk for autism among those who received vaccines that contained thiomersal.

There is no evidence relating autism either to the MMR vaccine or to thiomersal. Rather than consider the possibility that vaccinations might not cause autism, however, opponents of vaccinations offer new, speculative theories about what links them with autism. Over the last few decades, the number of childhood vaccinations that are recommended within many developed countries has increased significantly. As of 2014, the U.S. Centers for Disease Control and Prevention recommend that children be vaccinated against sixteen diseases. At certain milestones, an infant might receive as many as five injections in one healthcare visit. This has led some to worry that some infants' immune systems might not be strong enough to cope with the introduction of all the antigens that vaccines contain. If a child's immune system is weakened by the demands of current vaccination schedules, could this contribute to the onset of autism? The concern sounds reasonable, at least by the standards of common sense.

Health professionals, however, have at their disposal an array of evidence and arguments that reveal the concern is again not sensible. For example, although the number of recommended childhood vaccines has increased, improvements in relevant vaccine technologies and design mean that the total number of antigens contained across all recommended vaccination is significantly lower now than it has been in the past. Today's vaccine program places fewer demands on a child's immune system than previous programs. Second, studies reveal that vaccinated children are no more vulnerable than non-vaccinated children to those diseases for which vaccines are not available, a conclusion that undermines the idea that vaccinations weaken a child's immune system. The suggestion that vaccinations cause autism by weakening a child's immune system fares no better than those relating autism to the MMR vaccine or thiomersal.

Despite the absence of any reliable evidence, those who promote the autism-vaccination connection behave as if the connection is already established. They may admit that they don't know *how* vaccines cause autism, or that their understanding is at a rather speculative stage, but they are unwavering in their conviction that vaccinations are to blame. Previous failed attempts to explain the connection don't diminish their resolve. Wakefield pinned responsibility on the measles component of the MMR vaccine. Attempts to relate vaccines to autism via thiomersal, or weakened immune systems, have been equally effectively discredited. The absence of a convincing and well-researched underlying mechanism that would connect vaccinations with autism is not itself sufficient to reject the connection. Claims about causal relations often precede a deeper understanding of how one type of thing causes another. Nevertheless, the undeniable problem for those who attribute autism to vaccination is that they have never mustered any positive evidence for any of their claims. The particular hypotheses they have advanced have each failed, and hence those who oppose vaccinations must rely on magnifying uncertainty, casting suspicion on the health profession and pharmaceutical companies and appealing to parental fear and confusion.

Nicole Nattrass notes that the same four roles that she identified within the AIDS denialist community are also apparent among the anti-vaccine movement. Andrew Wakefield is the most conspicuous example of a hero-scientist. There are also examples of cultropreneurs, willing to sell expensive, ineffective and, in some cases, dangerous treatments to the parents of autistic children. Several journalists have aggressively promoted the idea that vaccinations cause autism. Documentaries have been released, designed for similar effect. Jenny McCarthy has become an important spokesperson for the movement and has appeared on numerous national chat shows. Nattrass further suggests that the parents of autistic children, who describe publicly the daily challenges of raising severely autistic children and who attribute the condition to vaccinations, function as the living icons. Their experiences and judgement are often valued more highly by other parents, than the conclusions of medical professionals.

As with AIDS denialism, the consequences of the anti-vaccination movement are severe. Outbreaks of diseases like mumps, measles and whooping cough are being recorded. Thousands of cases have been reported, some of which have resulted in death. Almost always, these deaths and outbreaks are avoidable. These are preventable diseases. It is rising mistrust of vaccinations, based on nothing more than rumour, speculation and ignorance, that is responsible for these tragedies. Turning the tide of public opinion admits of no easy answer. Relying on those who are most qualified to judge an issue, however, rather than those who create the most noise, would represent a very good start.

With both AIDS denialism and the anti-vaccine movement, it is abundantly clear that no genuine controversy exists among the relevant professional body. The overwhelming scientific consensus is that HIV causes AIDS, modern antiretrovirals are effective at managing the infection and vaccines don't cause autism. In 2000, more than five thousand physicians and scientists signed the Durban Declaration in response to events in South Africa. The paper described the evidence relating HIV to AIDS as 'clear-cut, exhaustive and unambiguous, meeting the highest standards of science' and continued, 'Drugs that block HIV replication… . Where available, reduced AIDS mortality by more than 80%'.[7] Among the signatories were eleven Nobel Prize winners and the directors of such notable scientific academies and societies as the Royal Society of London, Max Planck institutes, European Molecular Biology Organization, Pasteur Institute in Paris, U.S. National Academy of Sciences and U.S. Institute of Medicine. The World Health Organization, U.S.

[7] Durban Declaration (2000).

National Association of Sciences, U.S. Centers for Disease Control and UK National Health Service, among others, have investigated the link between vaccinations and autism, and have found no connection.

Surveys of gay and bisexual men, or of people who have HIV/AIDS, suggest that between 17 and 45 per cent of these demographics doubt that HIV causes AIDS, and similar proportions believe that HIV drugs do more harm than good.[8] Surveys likewise suggest that around one fifth of Americans believe that vaccines cause autism with another third being unsure. It is not uncommon to see reference made to the *MMR vaccine controversy* or the *thiomersal controversy*. Simply labelling these as controversies might lead us to suppose that there exists an actual controversy, as opposed to a well-publicized fear that has no basis in evidence, and that is driven by unwarranted suspicions, misplaced desires and no doubt a battery of cognitive biases. Among both AIDS denialists and the anti-vaccine movement, there exists significant suspicion of major pharmaceutical companies and a willingness to attribute the evidence within AIDS science or vaccine safety to global conspiracies. The relevant expert communities are in very little doubt surrounding the central claims under consideration.

The discrepancy between expert and public opinion suggests that these debates are created controversies. However, our three indicators of a created controversy do less to re-enforce that assessment. First, the independent and threatening agenda is not apparent in either case. Nattrass's cultropreneurs have a clear incentive to dissent from mainstream science, and in some cases we may find that individuals, like Andrew Wakefield, do benefit financially by opposing mainstream medicine. Those infected with HIV will understandably want to believe that HIV doesn't cause AIDS, parents of autistic children will often harbour a profound need to understand *how* their children developed autism and those scientists who have publicly announced their opposition to a given scientific conclusion may find it hard admitting that they were wrong. In general, however, there is no unifying agenda at work here analogous to the controversies created by the tobacco companies, for example.

Second, the motif of AIDS denialists and those who assert that vaccines cause autism seems better described as the ongoing advancement (and stubborn defence) of theses that have been discredited, rather than an attempt to promote uncertainty. Whether we're discussing climate change, Creationist science, the connection between cigarettes and lung cancer, AIDS denialism or the connection between vaccinations and autism, there is always a mix of magnifying uncertainty and promoting alternative (albeit quickly discredited) explanations. My impression of these issues, however, is that opponents to mainstream sciences in the latter two cases lean more towards insisting on their own conclusions rather than merely inflating uncertainty.[9] Our third indicator points more strongly towards a created controversy,

[8] Kalichman et al. (2010).

[9] An effect of promoting alternative hypotheses (and thereby denying mainstream scientific conclusions) will be to generate uncertainty, but this is different from *generating* uncertainty by *magnifying* uncertainty. AIDS denialists,

since AIDS denialists and the anti-vaccine movement are often deeply suspicious of the medical professions and thus less inclined to engage them. Efforts are directed far more towards convincing the public.

The fact that our three indicators might gesture only weakly towards the conclusion that these are examples of created controversies does nothing to reprieve either AIDS denialists or those who argue that vaccines cause autism. Such obstructionists have offered no reasonable objection to the scientific arguments they reject, and have no good response to the compelling evidence that scientists have accumulated. The absence of a unifying independent agenda and a tendency perhaps towards promoting alternative conclusions rather than magnifying uncertainty represent departures from the concept of a created controversy as it was introduced in earlier chapters. AIDS denialism and the anti-vaccine movement might be more appropriately labelled as conspiracy theories, rather than created controversies. But all this only illustrates an important, possible limit to the utility of our central concept.

Our concern is to hold reasonable beliefs. As part of this objective, we should recognize that the appearance of controversy surrounding some given conclusion might not represent a genuine controversy. We have discussed several clues that the appearance of controversy might be illusory. On occasions when these clues are not apparent, we shouldn't be assuming that all attitudes towards the central issue are reasonable. The cases discussed in this chapter suggest that the appearance of controversy can be misleading, even if no unifying agenda is present, and denial is more prevalent than uncertainty. Quibbling over the correct description (whether or not an issue deserves the title *created controversy*) is less important than reasonable evaluations of available evidence.

GMOs

In 2002, Zambia faced a severe food shortage, but the Zambian president refused a donation of 35,000 tons of American corn (maize) for fear that some of the corn had been genetically modified. Millions of Zambians faced starvation, but concerns with genetic modification outweighed this profound human need. Fears and uncertainties surrounding genetically modified crops are not restricted to sub-Saharan Africa. More than sixty countries (many industrialized, including all members of the European Union) have significant restrictions on the development and cultivation of genetically modified food, and some have complete bans. Many plant breeders and geneticists, molecular biologists and toxicologists regard such policies as unjustified, unnecessary and an obstacle to solving genuine, global problems.

I'm suggesting, are more invested in the former; climate change denialists, I suggest, more invested in the latter. These judgements are based on my own subjective impressions of large, complicated discussions. The differences might in reality be less pronounced.

Proponents argue that genetically modified organisms (GMOs) are safe for human consumption and better for the environment, that genetic modification will enable us to improve the flavour, safety and nutritional value of our food, while simultaneously reducing waste and protecting against diseases and other environmental threats that are capable of devastating entire crops. Critics of the technology argue that the risks to human health and the environment are not sufficiently well understood and potentially of enormous significance. Critics are suspicious of the companies that develop GMOs, and vocal in their support for policies that would enable consumers to avoid GMOs if they choose.

The stakes once again are high. In refusing the American corn, the Zambian president announced that 'we would rather starve than get something toxic'.[10] Some GMO research projects have seemingly been abandoned because, although developers were cautiously optimistic that they could improve certain food types, project funders suspected that consumers wouldn't want a genetically modified alternative, even if it was safer and healthier. Many proponents for GMOs argue that the mounting challenges associated with feeding the world population can be met only if we embrace the technology. If the attacks on GMOs lack merit, then we are losing out. On the other hand, hundreds of millions of acres are now being planted with GM crops, and upwards of 70 per cent of processed food in the United States contains GM ingredients. If GMOs are inherently unsafe, either to us or the environment, or are insufficiently regulated and serve principally to benefit global corporations, then our current practices are failing us and deserve censure.

Unfortunately for those looking to better educate themselves on the many issues surrounding GMOs, there is much misleading information and emotional rhetoric in circulation. There are competing sources for news, evidence and analysis, but it can be hard to know which ones to trust. GMO crops have been blamed for declining populations of monarch butterflies and for rising suicide rates among Indian farmers. The biggest producer of genetically modified seed is Monsanto, which has been accused of suing farmers for illegally growing Monsanto's GMOs on occasions when those GMOs were only introduced through wind-blown pollen from neighbouring farms. Suspicion surrounds biotechnology's habit of *patenting* particular genes, of contractually obligating farmers not to save seeds for subsequent seasons, of developing plants that would only produce sterile seeds and of manipulating regulatory agencies for favourable policies. The accuracy and significance of these stories has been challenged and, according to some commentators, in some cases thoroughly discredited, but critics of the technologies are often unpersuaded by these expositions. Advocates for GMO technology accuse their opponents of being anti-science. Critics of GMOs accuse the scientists and supporters of being industry shills. Here we can

[10] Quoted in Fedoroff and Brown (2006, 310).

only briefly review some of the core aspects of the discussion and will necessarily elide many details.

A good place to start our overview is by thinking about how GMOs are different from non-genetically modified food. As plant breeders and geneticists remind us, we have been modifying our crops for thousands of years. At first, this was achieved through selective breeding: the seeds from those plants which produced useful or desirable seeds or fruit were saved for planting in subsequent seasons. The seeds of some plants, however, grow into trees that produce fruit quite unlike the parent tree. Apple seeds are a good example. Cultivated apple trees don't come from seeds, but from grafting branches of a desirable apple tree onto the trunk and root system of another.

Hybridization is another important method by which modern crops have been developed. Hybridization mixes the genomes of both parents and produces something new. In a very small proportion of cases, this process will result in something better, in some respect, than either parent plant. Breeders have crossed millions of plants, often of different species, in their efforts to produce better, or just novel, fruits or seeds. In addition to mixing different genomes, plant breeders can also generate novelty by inducing genetic mutations in plant genomes. This practice started in the 1950s. Mutations have been induced using both radiation and chemicals. One chemical became particularly popular because it doubled the number of a plant's chromosomes. These and other methods not typically classified as genetic modification have given rise to many of the foods we unthinkingly consume. Sometimes these new foods and methods were greeted with suspicion and resistance, and were regarded as unnatural and potentially dangerous.

When people talk about GMOs, they are generally referring to methods that take a single gene, or perhaps a few genes, from one plant, animal, virus or bacterium, which is then introduced into another organism. Golden rice, for example, is a genetically modified food. Geneticists spent an enormous amount of time and energy trying to introduce into rice plants a gene that produces beta-carotene. Beta-carotene is a compound that animals and humans can convert into vitamin A. It is estimated that worldwide there are 250,000 children every year who become blind as a result of vitamin A deficiency and whose primary food source is rice. Introducing the beta-carotene into rice has potentially enormous social value. Some plants do produce the compound naturally. The challenge confronting scientists was to isolate the responsible genes, getting those genes to function within rice plants, but without compromising the viability of the plant or introducing environmental or health risks.

The production of GMOs is different from other ways of modifying crops. However, condemnation of all and only GMOs requires evidence that these differences render GMOs riskier than those methods of food production that involve genetic disruption via chemical and radiation-induced genetic mutations, doubling of chromosomes or creation of hybrids from different species. Food safety is itself a multifaceted, complex, imperfect collection of ideas, practices and policies which involves public demand, businesses that are trying to turn profits and government oversight that must balance health and environmental risks against economic considerations. Perhaps the most unfortunate feature of conversations surrounding GMOs is the implication that either they all share the same deeply troubling risk or they are all safe. In the 1990s, a genetically modified soybean was being developed. The intent was to increase its nutritional content by introducing a gene from the Brazil nut. The protein that was desired was not a known allergen at that time, but because Brazil nuts are an allergenic food, scientists tested the new soybean. These tests revealed that some people would be allergic to the soybean. It was withdrawn from production, as of course it should have been. Not all GMOs are sufficiently risk free to put on the market, but this provides very poor justification for prohibiting their development and distribution more generally.

It is important to recognize how distantly removed most of us are from the enormous challenges associated with growing, packaging and transporting food, and hence how unaware we are of the many methods and innovations that have been introduced to overcome these challenges. Crops can be lost to disease, insect infestation, drought and flooding. Farmers must attend to soil quality, weed control, immediate environmental impact, short- and long-term consequences of available practices and consumer demand, all while striving to ensure the ongoing sustainability of the business. We shouldn't surrender to the romantic notion that Mother Nature will relinquish all the sustenance and nutrition we need without significant coaxing and manipulation on our part. There's nothing natural about agriculture, and so we can't pretend that the problem with genetic modification is simply that it is unnatural.

Furthermore, we shouldn't assume that prohibiting the development of GMOs is risk free, and hence that any risks associated with the techniques are unnecessary. The production and consumption of *any* food introduces some risk. Many tragedies and disasters have hit the food industry that had nothing to do with genetically modified organisms. (A breakout of the E. coli bacterium in Germany originated at an organic spinach farm and is blamed for over thirty deaths; we don't conclude that modern organic farming is inherently dangerous and that its products should be avoided.) Furthermore, GMOs *might* provide an essential component in our battle to feed a growing global population amidst a changing climate. Of course, they might not, but unless we're sure that they don't, then restricting their development involves risk.

The central, public concern with GMOs is the risk they pose to human health. Polls attest that many consumers are either concerned about the safety to human health or are unsure about their safety. The scientific consensus, however, is that GMOs are as safe as the alternatives, and perhaps more so. The American Association for the Advancement of Science, World Health Organization, Royal Society of Medicine, American Medical Association, National Academy of Science, European Commission, Academy of Nutrition and Dietetics, American Society of Plant

Sciences, American Society for Cell Biology and many more learned societies and scientific bodies have concluded, and released statements asserting that GMOs are safe for human consumption. A 2013 analysis of the previous ten years of scientific research into the safety of GMOs reviewed almost 1800 scientific papers, many of which were publicly funded. The authors of the review article concluded that 'research conducted so far has not detected any significant hazards directly connected with the use of [genetically engineered] crops'.[11] Of course, the absence of evidence of risk, as we've noted, is not always evidence that no risk exists, but the number of studies that have been conducted make it reasonable to conclude that there is no general risk to human health associated with genetic modification.

One of the putative benefits of GMOs is that they're better for the environment. Advocates have suggested that GMOs reduce the need for pesticides, promote more efficient use of herbicides and help reduce soil erosion. By far the most widely grown GM crops are those that either produce their own insecticide or are resistant to certain herbicides. Bt corn, for example, has been engineered to produce a bacterium that is found naturally in soil and that produces proteins that are toxic to several species of insect. These proteins have no effect on humans and have been used as an insecticide for much of the twentieth century. Crops that produce the insecticides don't need spraying, which should both reduce the use of pesticides and restrict attention to those insects that threaten the crop, so beneficial insects survive.

With regard to herbicide use, varieties of soybean, corn and cotton have all been developed which are resistant to glyphosate. Glyphosate is the active ingredient in Monsanto's widely used herbicide Roundup, hence these GMOs have become popularly known as Roundup Ready. Farmers who plant these seeds can spray their fields with Roundup, knowing that the crop will remain unaffected but weeds which would damage the crop will be killed off. The products have proved enormously popular. So long as Roundup is effective against threatening weeds, farmers will only need one herbicide rather than a variety, and may turn the soil less for purposes of managing weeds.

It is of course important to distinguish the environmental benefits that have been promised from those that have actually been felt. It is just as important to recognize the environmental problems that GMOs have engendered. With regard to the latter, critics observe that the widespread use of Roundup-Ready crops and Bt corn has given rise to weeds that are resistant to glyphosate and insects that are Bt resistant. To critics, such developments are signs of the futility of attempting to impose our collective will on the natural world. To proponents, the emergence of insecticide and herbicide resistance is motivation to develop new products and new methods. Whether pesticide use has fallen since the introduction of GMOs is also contentious. The evidence for reduced insecticide use appears stronger. Assessing herbicide use

[11] Nicolia et al. (2013).

is further complicated by the importance of differentiating between more or less potent chemicals.

If we restrict our attention to questions of GMOs' safety and environmental impact, then the public outrage directed towards them would seem disproportionate to the amount of evidence of risk or harm. There is a strong scientific consensus that GMOs present no greater risk to human health than foods which are not genetically modified. Large-scale farming inevitably has environmental consequences, but if the cultivation of GMOs poses special risks, these have not been well evidenced.

Much of the discomfort with GMOs seems to originate in emotion rather than actual risk. As we noted in Chapter 5, public perceptions of risk don't always track actual risk. Paul Slovic's work on the psychology of risk assessment provides evidence that people judge as riskier that which is unfamiliar, involuntary, inequitably distributed and where lack of public information makes evaluating risk hard to evaluate.[12] GMOs score highly on all these dimensions, which would help explain public insistence that customers should be afforded the choice to avoid GMOs. The absence of credible evidence of actual risk, coupled with a plausible psychological explanation for public fear, suggest that controversy is created not genuine. However, there are additional factors to consider here, which complicate efforts to reliably evaluate the state of current debates.

Successfully engineering plants for desirable traits involves formidable technical challenges. Molecular geneticists must identify those genes that are responsible for the trait they wish to transfer from one organism to another, isolate the genes and, most difficult, then make them function within the new genome. These genes must continue to function outside the laboratory in field tests, and consumers must then be willing to buy the products. Early enthusiasm for biotechnology suggested that products would become available quickly, but experience has now taught us that decades of work can separate the conception of an idea for genetic modification and its availability to consumers, assuming success of any measure. The research costs associated with developing these products are daunting, and returns on those investments take a long time to materialize. Consequently, it is huge agricultural biotechnology corporations that dominate the world of genetic modification. The role of these corporations gives rise to important political and social questions concerning how the technologies of genetic modification are being used, for what purposes, who is ultimately going to benefit from these endeavours and how regulation of these products will be managed.

Marion Nestle argues that the most frequently voiced concerns of GMO critics concerning risks to human health and the environment are 'a surrogate for concerns about larger social issues'.[13] Concerns about safety and potential environmental risks are easy to promote, can quickly gain attention and support and can be hard to

[12] Slovic (1987). [13] Nestle (2003, 142).

suppress even if they lack any merit. Focusing public attention on these matters is an effective means of stimulating opposition to GMOs. However, for Nestle, the primary issues are about corporate greed, corporate political influence, conflicts of interest and failure of democratic processes. Even if the safety record of GMOs is uncontroversial and even if the environmental risks are negligible, there remain legitimate worries about their regulation.

Bioagricultural companies need to satisfy their investors, and so companies are motivated to pursue those products that promise the quickest and largest returns. These products may have little to do with feeding the world or fulfilling many of the benefits that genetic modification promised. Such companies are likewise motivated to lobby for policies and regulations that ease the transition from research and development to sales and export. They compete with activist groups to overturn state laws that might impact their profits and, due to their enormous financial clout, have generally been successful. Even if these products are safe, public trust will suffer if the purveyors are perceived to be manipulating the system. Sir Gordon Conway is supportive of biotechnology within agriculture, but is nevertheless critical of several aspects of the industry. In a 1999 speech, while serving as president of the Rockefeller Foundation, Conway described how 'The single biggest concern in the developing world may be that millions of poor farmers will become dependent on a dozen or so multinationals for the future livelihoods.'[14]

In a different context, ethicist Clive Hamilton contrasts Promethean and Soterian attitudes towards the Earth.[15] Prometheans are disposed to control and dominate, often by utilizing new technology and understanding. Soterians are more cautious and more concerned about the consequences of employing technology without sufficient sensitivity to the potentially devastating consequences that may result. The extent to which individuals identify more with Prometheans or Soterians will differ by degree, but this again helps us appreciate differing attitudes towards GMOs that have little to do with the accumulated evidence for risk to human health or the environment.

There are plenty of examples of environmental problems that were induced by human activity, through deliberate or accidental introduction of non-native species, intensive farming, resource depletion, not to mention accidents at nuclear power stations, oil and chemical spills and various other forms of ground, air and water pollution. We have a long history of creating problems through technological advancement; thus we can appreciate why some groups would prefer that we consider the implications of our practices far more closely before enacting processes that may be hard to reverse. Evidence of glyphosate-resistant weeds or Bt-resistant insects is for Prometheans simply a call to develop new technologies that will resolve the new problem. Insofar as agriculture is a struggle against nature to impose our will, nature's assaults on our

[14] Quoted in Kloppenburg (2005, 291).
[15] Hamilton (2013), which is actually a very nice book on the ethics of climate engineering.

ambitions are incentives to muster greater technological resources in opposition. For Soterians, the resistant weeds and insects are evidence for the futility of attempting to harness nature, and hence the importance of relinquishing ambitions of control and domination, and discovering less invasive methods for meeting human needs.

Those who are suspicious of rising corporate globalization, unchecked and deregulated free markets, international trade agreements that benefit large economies, efforts to control and manipulate biological and ecological systems, and who suppose that too many multinational corporations are insufficiently motivated to protect against social injustice and environmental disasters, may come to regard genetically modified organisms as the products of irresponsible sciences, of sciences that hurt rather than help society. These are concerns that have little to do with any immediate and direct threat to human health or environment, but they are serious concerns nonetheless. These are concerns which are harder to articulate, harder to understand and where the steps needed to address them are more difficult to appreciate. It is thus perhaps understandable that opposition to GMOs appears focused on issues of safety. Unfortunately, this focus undermines the epistemic merits of the cause.

For the reasons presented, whether the conversations surrounding GMOs constitute a genuine or created controversy is more complicated than those that involve vaccine safety and AIDS denialism. Those who urge that genetically engineered food is a risk to human health cherry-pick their data, magnify uncertainty, ignore the peer-reviewed literature, and conflate their mistrust of large corporations with evidence that the products are unsafe. There is a robust scientific consensus that GMOs present no special risk no human health. Nevertheless, there are legitimate concerns with the management of biotechnology and genetic modification that are harder to evaluate and should not be ignored or dismissed. These are criticisms of the relevant sciences, because critics are observing that these developments may not be serving public interests as well as they could, lack the kinds of oversight that we might deem appropriate, and engender an attitude towards our planet that many find objectionable. There are convoluted, important questions that we can only gesture towards here, but we still might hope to positively reorient discussions about the real risk and benefits of genetic modification.

Conclusions

Whether we're discussing issues of medicine and health, food safety, energy production, educational, social and foreign policies, the details surrounding important historical events or surprising scientific claims, we do well to remember that the appearance of controversy can be misleading, that subconscious fears, desires and pre-existing beliefs influence human reasoning, that sciences never establish their conclusions with certainty and that reaching reliable personal beliefs requires us to think carefully, and responsibly, to become as well informed as we can, and to pay close attention to the reliability of our sources.

Of vaccine safety and AIDS denialism, we saw in this chapter that our concept of a created controversy neither applies nor fails to apply unambiguously. Certainly there is no good reason to doubt the value of vaccinations, or the basic sciences of HIV and AIDS, but opponents are not bound by any common motivation, and less likely to magnify uncertainty rather than claim medical professionals are involved in a global conspiracy. These differences may persuade us that the label of created controversy doesn't apply to these issues, but there should be no temptation to suppose that we are dealing with a substantive, and genuine, scientific controversy. Of GMOs, however, a stronger case can be made that a genuine scientific controversy does exist. What distinguishes the case of GMOs from the other issues we've confronted in this chapter is to be understood in terms of values.

With vaccinations and AIDS, our objectives are straightforward. Our collective hope is to reduce, if not eliminate, the spread of terrible and potentially fatal diseases or, if diseases can't be cured, to prolong patients' lives and improve their quality of life. Health and medical professionals are unequivocal in their verdict that vaccinations and antiretrovirals are hugely successful, relative to these goals, and that the side-effects and risks are very minor by comparison. Of global food production, however, the objectives are more complicated. Certainly, there is a shared desire to ensure access to the food, water and nutrients that the global population requires. Similarly, most seem committed to achieving these ends via methods that are environmentally and economically sustainable. However, there are deep ideological divides over how these goals are best pursued, and this opens space for genuine scientific controversies. How do we properly balance Promethean against Soterian attitudes? How do we ensure that research, development and policy are benefiting the right groups? Are all stakeholders being heard? Are regulators being objective? The controversies that these questions induce concern neither safety nor environmental risk peculiar to GMOs, but instead the roles of sciences, technologies and values in shaping the modern world.

Discussion questions

1. To what extent do you think fear of vaccinations is a product of the appearance that there exists a genuine medical controversy? What other factors do you think contribute?
2. Do you think governments should require healthy individuals to receive vaccinations, if not being vaccinated puts others' health at risk?
3. The mechanisms by which vaccinations were thought to cause autism have each been thoroughly discredited. How does this weaken the general hypothesis that vaccinations cause autism?
4. Suppose that some readers of this book had not previously heard of AIDS denialism or the discredited connection between vaccinations and autism. Are they likely to attach more

credence to these ideas than previously, merely by being exposed to them? Are there ways to avoid this consequence?

5. Does it matter whether we come to regard AIDS denialism or the suggestion that vaccines cause autism as created controversies?
6. Many people would like foods to be labelled, so consumers can see whether it contains, or may contain, genetically modified ingredients. What do you think are the strengths and weaknesses of this proposal?
7. Why might we worry about large bioagricultural companies achieving ever more influence within global agricultural policymaking and research development?

Suggested reading

For more on AIDS denialism, see Kalichman (2009) and Nattrass (2012). Offit (2010) and Mnookin (2012) are very useful resources for understanding the discredited connection between autism and vaccinations. For the sciences of GMOs, see Fedoroff and Brown (2006) and Stewart (2003), and for the history and politics, see Nestle (2003). Online, a worthwhile series of articles on GMOs can be found at grist.org/series/panic-free-gmos/.

Points to remember: Part III

1. A created controversy is an issue on which the overwhelming majority of experts are in agreement, but the wider public supposes that the issue is contested and the relevant scientific community evenly divided.
2. Insofar as the appearance of controversy might lead people to suppose that no-one really has a good grasp on certain issues, we are in danger of ignoring important information.
3. Ignoring important information threatens the reliability of individual choice and sound social policy.
4. There is no controversy surrounding the core tenets of anthropogenic climate change.
5. There is no controversy concerning the age of the Earth, the core assumptions of biological evolution, nor the acute weaknesses of ID.
6. It is uncontroversial that HIV causes AIDS, and vaccines don't cause autism.
7. Any impression that any of these issues represent genuine scientific controversies speaks principally to the success of those groups that have created the appearance of controversy.
8. There is no evidence to suggest that GMOs present a special threat to human health. There are, however, legitimate worries concerning who benefits from the technology and whether it is being judiciously utilized and regulated.
9. The idea that scientific communities are involved in global conspiracies is implausible.
10. Our own uncertainty about certain issues is very likely a poor guide to the current state of many scientific issues, and hence a poor reason to reject scientific consensus. The layperson typically has an extremely meagre sense for the sophistication, rigour, scale and variety of modern scientific research programs, the range of methods utilized to gather data, the extraordinary number of very narrow questions being posed, the degree of coherence between distinct methods and analyses, and the depth of understanding that scientists have achieved.
11. There remain many interesting important questions involving the changing climate and our responses, the relationship between science and religion, biological evolution and related disciplines, issues of public health, the costs and benefits associated with new technologies, and so on. These unresolved and important questions shouldn't detract from what scientific disciplines have achieved.

12. Scientific methods, conclusions and goals are always legitimate objects of criticism and analysis. However, if we don't work hard to distinguish legitimate objections from weak and politically or ideologically motivated objections then we risk a great deal.

Concluding remarks

When significant scientific consensus is concealed and the impression is created that particular claims are controversial within the relevant scientific community, dangerous and regrettable consequences follow. First, we produce a population that is less well informed that it should be. For all our cognitive failings and limitations, we are capable of grasping and understanding remarkable and astonishing features of our world, ourselves, our past, our co-inhabitants on this plant, living cells and the molecular machinery that drives them, distant galaxies and stars, and the forces that move, shape, alter and maintain everything from the smallest constituents of matter to the fabric of the cosmos. Not everyone cares to invest significant time and energies tracking the moving frontiers of modern scientific investigation, but general scientific acumen would no doubt be improved if we were better at distinguishing worthwhile criticisms of modern scientific conclusions from frivolous ones, if we learned to place a more appropriate level of confidence in our own abilities to evaluate complex issues and if we better appreciated both the unavoidable fallibility of scientific conclusions and yet also the complete and unambiguous compatibility between scientific fallibility and the existence of overwhelming evidence that, concerning particular scientific claims, we almost certainly are not wrong.

Levels of scientific literacy lower than they otherwise would be is one regrettable consequence of creating controversy. More practical consequences are also apparent. When we ignore expert opinion, we risk our health, our well-being and the quality of life of future generations. We should be vigilant and alert to the possibility that something might appear controversial and uncertain only because that appearance benefits a select few at the expense of a great many. Once we are suspicious that the appearance of controversy has been created, we should be extremely cautious about rejecting the conclusions of scientific consensus. We should be wary that our emotions, desires, overconfidence and personal experiences are playing an irrational and central role in our reasoning and judgement. We should seek the most reliable evidence, rather than settle for what's easiest to recall. We should evaluate arguments critically and consider the possibility that better explanations for cited evidence might be available but are unknown to us personally.

For all the ways in which we can improve our reasoning and evaluations, however, it would be naïve to suppose that the ideas and tools described in this book can cure all the ills that created controversies induce. Psychologists and sociologists have gathered evidence suggesting that simply educating people better on the relevant sciences often isn't enough. Revealing the flaws in particular arguments doesn't always suffice. When scientific conclusions are interpreted as having non-trivial implications for commerce, agriculture, global corporations, industries and the tens of thousands they employ, prevalent ideologies and personal liberties, then we can appreciate why the range of responses would be varied, entangled, convoluted and exhausting. People react differently to complex issues, depending on personal experience, age, gender, race and socio-economic background. When individuals are directly and personally affected, then it is entirely appreciable that they would become less inclined simply to defer to scientific authority. Parents want to better understand why they should be persuaded that vaccines are safe. Curious citizens want to know why particular studies have been dismissed as worthless, why particular objections to established scientific opinion are dismissed as unintelligible, how new technologies are being evaluated, and by whom, whether executive committees and panels are being as objective as they should, who makes up those groups and what agenda they may have.

There are important questions about how scientific communities should respond to genuine and serious inquiry. Responding to inquisitive laypersons with ridicule and derision will be unproductive, but when high-profile commentators prove themselves incapable of rational argument, insensitive to compelling empirical evidence and seemingly unwavering in their support for ideas that have been clearly discredited, then it is understandable that their interlocutors can become impatient and irritated. Furthermore, as we noted with regard to GMOs, what separates the two sides of some debates might only superficially concern the empirical evidence. Perhaps values are, consciously or subconsciously, playing important roles within many discussions. As we noted, Oreskes and Conway describe how a small number of scientists added significant weight to sceptical causes, despite lacking relevant expertise, because they held a particular view about the role of government, a role that seemingly involved protecting the economic interests of major corporations no matter what costs were incurred by society at large. To the extent that debates are perpetuated by varying ideas about who sciences and technologies should serve, and how, we may never make progress unless these values are made explicit and openly discussed.

In many cases, the role of values within scientific discourse assumes particular urgency because there is a breakdown of trust. We noted that AIDS denialists and members of the anti-vaccine movement are distrustful of large pharmaceutical companies, and to some extent this attitude is understandable. Those who are suspicious of Monsanto seem similarly justified. Big business, including multinational pharmaceutical, oil and bioagricultural companies, are prone to cutting corners, covering up errors, exploiting legal loop holes and lobbying for favourable policy, all in the pursuit of a bigger market share. Insofar as scientific communities have aligned themselves with such companies, or have adopted questionable research habits more generally, they may need to work hard to regain the trust of the public. Groups that historically have been victims of greater oppression, subjugation and exploitation are more likely to be distrustful of certain kinds of authority.

Historians describe the twentieth and early twenty-first centuries as having experienced a profound growth in anti-intellectualism, particularly in the United States. There are several causes of this ascendancy, but the overall effect has been a degradation in the perceived value of education, academic qualifications, reasoned argument and expertise. Hopes of improving public attitudes towards scientific reasoning are clearly compromised by the more general poor regard in which expertise is held.[1]

The complexity of the social problems that orbit certain scientific questions, the role of values within at least some of these debates, problems of trust and suspicion of higher education all suggest that progress in the struggle against created controversies will require more than we have discussed in this book. Nevertheless, there is enormous benefit to developing more sophisticated attitudes towards scientific conclusions. Scientific controversies can be artificially created, and these threaten our prosperity, but if we are patient, judicious and respectful, eventually created controversies can be dismantled. Their destruction might be harder than their creation, but rather than cause for despair this should serve only as motivation to take greater personal responsibility in our attitudes towards issues of profound social importance, our evaluations and our willingness to discuss them wisely, honestly and intelligently.

[1] For an excellent history of anti-intellectualism, see Jacoby (2009).

References

Alexander, D., and R. Numbers. 2010. *Biology and ideology from Descartes to Dawkins* (Chicago: University of Chicago Press).

Anderegg,W., J. Prall, J. Harold and S. Schneider. 2010. 'Expert credibility in climate change', *Proceedings of the National Academy of Sciences*, 107, 12107–12109.

Archer, D., and S. Rahmstorf. 2010. *The climate crisis: an introductory guide to climate change* (Cambridge: Cambridge University Press).

Ariely, D. 2010. *Predictably irrational*, expanded revised edition (New York: Harper Perennial).

Bacon, F. 1999. *Selected philosophical works*, Hackett Classics (Indianapolis, IN: Hackett). First published in F. Bacon, *Novum Organum Scientiarum*, 1620.

Barbour, I. 1990. *Religion in an age of science*, Gifford Lectures 1989–1991 (New York: HarperCollins).

Barker, G., and P. Kitcher. 2013. *Philosophy of science: a new introduction*, Fundamentals of philosophy (Oxford: Oxford University Press).

Behe, M. 1996. *Darwin's black box: the biochemical challenge to evolution* (New York: Free Press).

Behe, M. 2007. *The edge of evolution: the search for the limits of Darwinism* (New York: Simon and Schuster).

Bekelman, J., Y. Li and C. Gross. 2003. 'Scope and impact of financial conflicts of interest in biomedical research: a systematic review', *Journal of the American Medical Association*, 289, 454–465.

Bird, A. 2001. *Thomas Kuhn*, Philosophy Now (Princeton: Princeton University Press).

Black, M. 1954. 'The inductive support of inductive rules', in *Problems of analysis* (Ithaca, New York: Cornell University Press).

Bloor, D. 1991. *Knowledge and social imagery* (Chicago: University of Chicago Press).

Bogen, J. 2014. 'Theory and observation in science', *The Stanford Encyclopedia of Philosophy* (summer edition), Edward N. Zalta (ed.), http://plato.stanford.edu/archives/sum2014/entries/science-theory-observation/.

Bowler P., and I. Morus. 2005. *Making modern science: a historical survey* (Chicago: University of Chicago Press).

Brandt, A. 2009. *The cigarette century: the rise, fall, and deadly persistence of the product that defined America* (Cambridge, MA: Basic Books).

Brockman, J. (ed.). 2006. *Intelligent thought: science versus the intelligent design movement* (New York: Vintage).

Cairns-Smith, A. G. 1990. *Seven clues to the origin of life: a scientific detective story* (Cambridge: Cambridge University Press).

Carroll, S. 2007. *The making of the fittest: DNA and the ultimate forensic record of evolution* (New York: W.W. Norton).

Ceccarelli, L. 2011. 'Manufactured scientific controversy: science, rhetoric, and public debate', *Rhetoric & Public Affairs*, 14, 195–228.

Chakravartty, A. 2008. 'What you don't know can't hurt you: realism and the unconceived', *Philosophical Studies*, 137, 149–158.

Chalmers, A.F. 2013. *What is this thing called science?*, 4th edition (Indianapolis, IN: Hackett Publishing).

Collins, H., and T. Pinch. 2002. *The Golem at large: what you should know about technology*, Canto (Cambridge: Cambridge University Press).

Cook, J., D. Nuccitelli, S. Green, M. Richardson, B. Winkler, R. Painting, R. Way, P. Jacobs and A. Skuce. 2013. 'Quantifying the consensus on anthropogenic global warming in the scientific literature', *Environmental Research Letters*, 8, 024024.

Coyne, J. 2009. *Why evolution is true* (New York: Viking Penguin Books).

Darwin, C. 2004. *On the origin of species* (New York: CRW Publishing). First published in 1859.

Dawkins, R. 2010. *The greatest show on earth: the evidence for evolution*, reprint edition (New York: Free Press).

Dembski, W. 2002. *Intelligent design: the bridge between science & theology* (Downers Grove, IL: IVP Academic).

Dixon, T. 2008. *Science and religion: a very short introduction* (Oxford: Oxford University Press).

Douglas, H. 2009. *Science, policy and the value-free ideal* (Pittsburgh: University of Pittsburgh Press).

Dupré, J. 1995. *The disorder of things: metaphysical foundations of the disunity of science* (Cambridge, MA: Harvard University Press).

Durban Declaration. 2000. *Nature*, 406, 15–16.

Elliott, K., and D. Resnik. 2015. 'Scientific Reproducibility, Human Error, and Public Policy', *BioScience*, 65, 5–6.

Fahrbach, L. 2011. 'How the growth of science ends theory change', *Synthese*, 180, 139–155.

Fedoroff, N., and N. Brown. 2006. *Mendel in the kitchen: a scientist's view of genetically modified food* (Washington, DC: Joseph Henry Press).

Ferngren, G.B. 2002. *Science and religion: a historical introduction* (Baltimore: Johns Hopkins University Press).

Flannery, T. 2001. *The weather makers: how man is changing the climate and what it means for life on earth* (New York: Grove Press).

Fodor, J. 1984. 'Observation reconsidered', *Philosophy of Science*, 51, 23–43.

Forrest, B., and P. Gross. 2004. *Creationism's Trojan horse: the wedge of intelligent design* (Oxford: Oxford University Press).

Franklin, A., A.W.F. Edwards, D.J. Fairbanks and D.L. Hartl. 2008. *Ending the Mendel-Fisher controversy* (Pittsburgh: University of Pittsburgh Press).

Futuyama, D.J. 2013. *Evolution*, 3rd edition (Sunderland, MA: Sinauer Associates).

Gelman, S. 2004. 'Psychological essentialism in children', *Trends in cognitive sciences*, 8, 404–409.

Gigerenzer, G. 2006. 'Out of the frying pan into the fire: behavioral reactions to terrorist attacks', *Risk Analysis*, 26, 347–351.

Godfrey-Smith, P. 2003. *Theory and reality: an introduction to the philosophy of science*, Science and its conceptual foundations series (Chicago: University of Chicago Press).

Godfrey-Smith, P. 2008. 'Recurrent, transient underdetermination and the glass half-full', *Philosophical Studies*, 137, 141–148.

Goldacre, B. 2010. *Bad science: quacks, hacks, and big pharma flacks* (New York: Faber and Faber).

Goldman, A. 2001. 'Experts: which ones should you trust?', *Philosophy and Phenomenological Research*, 63, 85–110.

Gould, S.J. 1997. 'Nonoverlapping magisteria', *Natural History*, 106, 16–22.

Hacking, I. 1983. *Representing and intervening: introductory topics in the philosophy of natural science* (Cambridge: Cambridge University Press).

Hacking, I. 1992. 'The self-vindication of the laboratory sciences', in Andrew Pickering (ed.), *Science as practice and culture* (Chicago: University of Chicago Press), 29–64.

Hacking, I., 2000. *The social construction of what?* (Cambridge, MA: Harvard University Press).

Hamilton, C. 2013. *Earthmasters: the dawn of the age of climate engineering* (New Haven, CT:Yale University Press).

Hanson, N.R. 1958. *Patterns of discovery: an inquiry into the conceptual foundations of science* (New York: Cambridge University Press).

Heath, C., and D. Heath. 2007. *Made to stick: why some ideas survive and others die* (New York: Random House).

Henig, R. 2001. *The monk in the garden: the lost and found genius of Gregor Mendel, the father of genetics* (Boston: Houghton Mifflin Harcourt).

Houghton, J. 2009. *Global warming: the complete briefing*, 4th edition (New York: Cambridge University Press).

Hoyningen-Huene, P. 1993. *Reconstructing scientific revolutions: Thomas S. Kuhn's philosophy of science* (Chicago: University of Chicago Press).

Hull, D. 1988. *Science as a process: an evolutionary account of the social and conceptual development of science,* Science and its conceptual foundations series (Chicago: University of Chicago Press).

Hume, D. 2000. *Treatise of human nature*, D. Norton and M. Norton (eds.), Oxford philosophical texts (New York: Oxford University Press). First published 1738.

Hurley, P. 2014. *A concise introduction to logic* (Boston: Cengage Learning).

Isaak, M. 2007. *The counter-creationism handbook* (Oakland: University of California Press).

Jacoby, S. 2009. *The Age of American Unreason* (New York: Vintage Books).

Johnson, P. 2002. *The wedge of truth: splitting the foundations of naturalism* (Downer's Grove, IL: IVP Press).

Kahneman, D. 2013. *Thinking, fast and slow* (New York: Farrar, Straus and Giroux).

Kalichman, S. 2009. *Denying AIDS: conspiracy theories, pseudoscience, and human tragedy* (New York: Springer,).

Kalichman, S., L. Eaton and C. Cherry. 2010. '"There is no Proof that HIV causes AIDS": AIDS denialism beliefs among people living with HIV/AIDS', *Journal for Behavioral Medicine*, 33, 432–440.

Kitcher, P. 1993. *The advancement of science* (New York: Oxford University Press).

Kitcher, P. 2003. *Science, truth, and democracy*, Oxford studies in the philosophy of science (Oxford University Press,).

Klein, P. 2008. 'Contemporary responses to Agrippa's trilemma', in *The Oxford handbook of skepticism* (ed.) J. Greco (New York: Oxford University Press).

Kloppenburg, J.R., Jr., 2005. *First the seed: the political economy of plant biotechnology*, 2nd edition (Madison: University of Wisconsin Press).

Kuhn, T.S. 2012. *The structure of scientific revolutions*, 4th edition (Chicago: University of Chicago Press). First published 1962.

Ladyman, J. 2001. *Understanding philosophy of science* (London: Routledge).

Lakatos, I. 1978. *The methodology of scientific research programmes: volume 1, Philosophical papers* (Cambridge: Cambridge University Press).

Latour, B., and S. Woolgar. 2013. *Laboratory life: the construction of scientific facts* (Princeton, NJ: Princeton University Press). First published 1979.
Laudan, L. 1981. 'A confutation of convergent realism', *Philosophy of Science*, 48, 19–49.
Laudan, L. 1983. 'The demise of the demarcation problem', in R.S. Cohen and L. Laudan, *Physics, philosophy and psychoanalysis: essays in honor of Adolf Grünbaum*, Boston Studies in the Philosophy of Science (Dordrecht: D. Reidel, 76), 111–127.
Leiserowitz, A., E. Maibach, C. Roser-Renouf and N. Smith. 2011. *Climate change in the American mind: Americans' global warming beliefs and attitudes in May 2011* (Yale University and George Mason University, Yale Project on Climate Change Communication).
Loftus, E. 1977. 'Shifting human color memory', *Memory & Cognition*, 5, 696–699.
Longino, H. 1990. *Science as social knowledge* (Princeton, NJ: Princeton University Press).
Magnus, P.D. 2010. 'Inductions, red herrings, and the best explanation for the mixed record of science', *British Journal for the Philosophy of Science*, 61, 803–819.
Maslin, M. 2009. *Global warming: a very short introduction*, 2nd edition (New York: Oxford University Press).
Miller, K.R. 2008. *Only a theory: evolution and the battle for America's soul* (New York: Viking Penguin Books).
Mnookin, S. 2012. *The panic virus: the true story behind the vaccine-autism controversy* (New York: Simon & Schuster).
Mooney, C. 2006. *The Republican war on science* (Cambridge, MA: Basic Books).
Nattrass, N. 2012. *The AIDS conspiracy: science fights back* (New York: Columbia University Press).
Nestle, M. 2003. *Safe food: the politics of food safety* (Oakland: University of California Press).
Nicolia, A., A. Manzo, F. Veronesi and D. Rosellini. 2013. 'An overview of the last 10 years of genetically engineered crop safety research', *Critical Reviews in Biotechnology*, 34, 77–88.
Nilsson, D., and S. Pelger. 1994. 'A pessimistic estimate of the time required for an eye to evolve', *Proceedings of the Royal Society of London*, Series B: Biological Sciences, 256, 53–58.
Norton, J. 2003. 'A material theory of induction', *Philosophy of Science*, 70, 647–670.
Numbers, R.L. 2006. *The creationists: from scientific creationism to intelligent design*, Expanded edition (Cambridge, MA: Harvard University Press).
Offit, P.A. 2010. *Autism's false prophets: bad science, risky medicine, and the search for a cure* (New York: Columbia University Press).
Okasha, S. 2002. *Philosophy of science: a very short introduction* (New York: Oxford University Press).
Okasha, S. 2005. 'Does Hume's argument against induction rest on a quantifier-shift fallacy?', *Proceedings of the Aristotelian Society*, 105, 253–271.
Oreskes, N. 2004. 'The scientific consensus on climate change', *Science*, 306, 1686.
Oreskes, N., and E. Conway. 2011. *Merchants of doubt: how a handful of scientists obscured the truth on issues from tobacco smoke to global warming* (London: Bloomsbury Press).
Otto, S. 2011. *Fool me twice: fighting the assault on science in America* (Emmaus, PA: Rodale Books).
Pigliucci, M., and M. Boudry. 2013. *Philosophy of pseudoscience: reconsidering the demarcation problem* (Chicago: University of Chicago Press).
Pilkey, O., and K. Pilkey. 2011. *Global climate change: a primer* (Durham, NC: Duke University Press Books).
Proctor, R. 2012. *Golden holocaust: origins of the cigarette catastrophe and the case for abolition* (Oakland, CA: University of California Press).

Pronin, E., D. Lin and L. Ross. 2002. 'The bias blind spot: perceptions of bias in self versus others', *Personality and Social Psychology Bulletin*, 28, 369–381.

Pronin, E., T. Gilovich and L. Ross. 2004. 'Objectivity in the eye of the beholder: divergent perceptions of bias in self versus others', *Psychological Review*, 111, 781–799.

Psillos, S. 1999. *Scientific realism: how science tracks truth*, Philosophical issues in science (London and New York: Routledge).

Quine, W.V.O. 1951. 'Two dogmas of empiricism', *Philosophical Review*, 60, 20–43.

Rahmstorf, S., G. Foster and A. Cazenave. 2012. 'Comparing climate projections to observations up to 2011', *Environmental Research Letters*, 7, 044035.

Ruse, M., and R. Pennock. 2008. *But is it science? The philosophical question in the creation/evolution controversy*, updated edition (New York: Prometheus Books).

Russell, B. 1999. *The problems of philosophy* (New York: Dover). Originally published in 1912.

Salmon, M.H. 2006. *Introduction to logic and critical thinking*, 5th edition (Belmont, CA: Wadsworth).

Salmon, W. 1967. *Foundations of scientific inference* (Pittsburgh, PA: University of Pittsburgh Press).

Sarkar, S. 2007. *Doubting Darwin: creationist designs on evolution* (Malden, MA: Wiley-Blackwell).

Schick, T., and L. Vaughn. 2013. *How to think about weird things: critical thinking for a new age*, 7th edition (New York: McGraw-Hill).

Schwab, I. 2011. *Evolution's witness: how eyes evolved* (New York: Oxford University Press).

Schwarz, N., I. Skurnik, C. Yoon and D. Park. 2005. 'How warnings about false claims become recommendations', *Journal of Consumer Research*, 31, 713–724.

Scott, E.C. 2009. *Evolution versus creationism: an introduction*, 2nd edition (Oakland: University of California Press).

Shanks, N. 2007. *God, the devil, and Darwin: a critique of intelligent design theory* (New York, Oxford University Press).

Shermer, M. 2012. *The believing brain: from ghosts and gods to politics and conspiracies – how we construct beliefs and reinforce them as truths* (New York: St. Martin's Griffin).

Shome, D., and S. Marx. 2009. *The psychology of climate change communication: a guide for scientists, journalists, educators, political aides, and the interested public* (New York: Center for Research on Environmental Decisions).

Shubin, N. 2009. *Your inner fish: a journey into the 3.5-billion-year history of the human body* (New York: Vintage).

Shulman, S., et al. 2012. *Cooler smarter: practical steps for low-carbon living* (Washington, DC: Island Press).

Simmons, N., K. Seymour, J. Habersetzer and G. Gunnell. 2008. 'Primitive Early Eocene bat from Wyoming and the evolution of flight and echolocation', *Nature*, 451, 818–821.

Slovic, P. 1987. 'Perception of risk', *Science*, 236, 280–285.

Sober, E. 1988. *Reconstructing the past* (Cambridge, MA: MIT Press).

Solomon, M. 2001. *Social empiricism* (Cambridge, MA: Bradford Books).

Stanford, P.K. 2006. *Exceeding our grasp: science, history and the problem of unconceived alternatives* (New York: Oxford University Press).

Stanford, P.K. 2013. 'Underdetermination of scientific theory', *The Stanford Encyclopedia of Philosophy* (Winter Edition), Edward N. Zalta (ed.), URL = <http://plato.stanford.edu/archives/win2013/entries/scientific-underdetermination/>.

Stewart, C.N. 2003. *Genetically modified planet: environmental impacts of genetically engineered plants* (New York: Oxford University Press).

Sutherland, S. 2007. *Irrationality*, 2nd edition (London: Pinter & Martin Ltd).

Svenson, O. 1981. 'Are we all less risky and more skillful than our fellow drivers?', *Acta Psychologica*, 47, 143–148.

Union of Concerned Scientists. 2012. *A climate of corporate control: how corporations have influenced the dialogue on climate science and policy* (Cambridge, MA: Union of Concerned Scientists).

Vickers, J. 2014. 'The problem of induction', *The Stanford Encyclopedia of Philosophy* (Fall Edition), Edward N. Zalta (ed.), URL = <http://plato.stanford.edu/archives/fall2014/entries/induction-problem/>.

Weart, S. 2008. *The discovery of global warming: revised and expanded edition* (Cambridge, MA: Harvard University Press).

Weaver, K., S. Garcia, N. Schwarz and D. Miller. 2007. 'Inferring the popularity of an opinion from its familiarity: a repetitive voice can sound like a chorus', *Journal of Personality and Social Psychology*, 92, 821–833.

Weber, M. 2009. 'The crux of crucial experiments: Duhem's problems and inference to the best explanation', *British Journal for Philosophy of Science*, 60, 19–49.

Whyte, J. 2004. *Crimes against logic: exposing the bogus arguments of politicians, priests, journalists, and other serial offenders* (New York: McGraw Hill).

Zammito, J.H. 2004. *A nice derangement of epistemes: post-positivism in the study of science from Quine to Latour* (Chicago: University of Chicago Press).

Index